Mary Anne

AF709943

Winnebago Cooperative Telephone Association History: 1950–1990

WINNEBAGO
COOP
TELEPHONE
ASS'N

Winnebago Cooperative Telephone Association History: 1950–1990

Robert M. Merryman

40th Anniversary Edition

Winnebago Cooperative Telephone Association
Lake Mills, Iowa

International Standard Book Number: 0-9627097-0-0
Library of Congress Catalog Card Number: 90-070843
© 1990 by the Winnebago Cooperative Telephone Association,
Lake Mills, Iowa, All Rights Reserved
Typed by Winnebago Cooperative Telephone Association
Printed by Stoyles Graphic Services, Lake Mills, Iowa, U.S.A.
Typeface is Palatino

Contents

Foreword

This *History of the Winnebago Cooperative Telephone Association* tells the story of a remarkable telephone company—its birth and growth. The contents have been carefully researched, documented and organized by the author.

By the year 1949, the Winnebago Rural Electric Cooperative had been in existence for ten years. At this point, the government began to encourage the development of telephone service for rural America. Glenn Bergland and his associates on the board of the Winnebago REC had the vision to recognize that better telephone service was overdue for rural communities throughout the nation. It was time to explore the possibilities for North Iowa. This they did, and through their dedication and hard work from Day One their new telephone association took root. Mr. Bergland formulated a plan that made the new telephone cooperative eligible for a federal loan.

Like every newly conceived idea, there were problems and obstacles to be overcome. This book reveals the dedicated effort these pioneers put forth to bring about a whole new vista in rural telephone communication. As a result of this dedication, the Winnebago RTA became a national model for other newly organized groups. They, likewise, were interested in bringing modern telephone service to their areas.

With the government loan finally awarded to the RTA, it was time to begin construction. Preliminary plans had been studied and authorized.

Mr. Bergland was a stickler for quality, an attribute that characterizes the management of the RTA to this day. The rundown systems of the old companies, which had been consolidated into the RTA, were brought up to standard. The old magneto phones were replaced with modern dial phones. The Winnebago RTA was off to an impressive start. This was followed by replacing poles and overhead wires and party lines with a network of underground cables and single party lines.

As you read the book, you will note that with time moving on, there are changes in personnel. There have been many dedicated persons serving on the board of directors over the years. Fortunately, there have been only two managers: Glenn Bergland, twenty-five years, and Ken Lein, sixteen years to date.

Mr. Lein has continued to increase the Cooperative's geographic area and reinforce its position on the ever advancing frontier of communications technology. Fiber optics, custom calling features, data banks, cable television, a choice of long distance carriers are all part of today's Winnebago RTA.

As the Winnebago Cooperative Telephone Association marks its Fortieth Anniversary, it remains a model of which all of us can be proud.

G. W. Aasgaard
Publisher, Lake Mills Graphic, Ret.

Preface

When Ken Lein asked me if I'd like to write the history of the Winnebago RTA, I jumped at the chance to tell the story of this remarkable telephone cooperative. I knew a little of how a kaleidoscopic assortment of several dozen small "whoop and holler" telephone companies had been consolidated into today's exemplary telephone entity; and I was aware of how we, the fortunate RTA subscribers, all too often accept as commonplace the RTA's advanced technology (including features that thousands of other telephone systems across the nation either don't offer or charge a premium for).

Really, I'm not the author of this history. I am, if you will, an amanuensis, meaning 'one who copies what another has written,' the secretary who has assembled a narrative from the records of the RTA, newspaper files, and the detailed notes of Glenn Bergland and Ken Lein, the two skippers of the RTA: Glenn, who, from the beginning and through the earlier years, put it all together, transforming the pole and wire companies into a single, modern, underground telephone system; and Ken, his successor, who has continued to expand the Co-op in its role of state-of-the-art telephone and cable TV company. *June 26, 1990*

Chronology of the Board of Directors

Winnebago Cooperative Telephone Association

The original board members upon the establishment of the Cooperative in January, 1950

Elmer O. Bergland, President

R. E. Aukes

M. J. Madson, Vice President

H. D. Hartman, Secretary-Treasurer

Archie Anderson

Gustaf Gustafson

M. J. Rygh

Over the ensuing years as these people vacated their positions, the following replacements occurred:

March, 1953:	Elmer Seim was elected to fill the vacancy created by the resignation of Elmer O. Bergland.
December, 1956:	Arlo Carson was appointed to fill the vacancy created by the death of M. J. Rygh.
January, 1958:	Norman P. Larson was appointed to fill the vacancy created by the death of H. D. Hartman.
February, 1966:	Vernon Jacobson was appointed to fill the vacancy created by the death of Archie Anderson.
January, 1967:	Victor Knudtson was appointed to fill the vacancy created by the resignation of M. J. Madson.
March, 1967:	DeVere Steffensen was appointed to fill the vacancy created by the death of Elmer Seim.
January, 1969:	Clarence Willand was appointed to fill the vacancy created by the resignation of Norman P. Larson.
March, 1971:	Milford Helgeson was elected to fill the vacancy created by the resignation of Clarence Willand.
January, 1973:	Lloyd Kronemann was appointed to fill the vacancy created by the resignation of R. E. Aukes.
January, 1975:	James Petersburg was appointed to fill the vacancy created by the resignation of Gustaf Gustafson.
September, 1979:	Keith Finer was elected to fill the vacancy created by the resignation of Arlo B. Carson.
May, 1985:	Jerry Epland was appointed to fill the vacancy created by the resignation of Victor Knudtson.

Presidents

1950-1953	Elmer O. Bergland
1953-1973	R. E. Aukes
1973-1974	Gustaf Gustafson
1974-1976	Arlo Carson
1976-Present	Vernon C. Jacobson

Vice Presidents

1950-1966	M. J. Madson
1966-1973	Gustaf Gustafson
1974-1976	Vernon C. Jacobson
1976-1985	Victor Knudtson
1985-Present	Milford Helgeson

Secretary-Treasurer

1950-1957	H. D. Hartman
1957-1966	Gustaf Gustafson
1966-1973	Arlo Carson
1973-1976	Victor Knudtson
1976-Present	Lloyd Kronemann

Board of Directors 1990

Vernon C. Jacobson, President

Milford Helgeson, Vice President

Lloyd Kronemann, Secretary-Treasurer

DeVere Steffensen

James Petersburg

Keith Finer

Jerry Epland

"The business and affairs of the Cooperative shall be managed by a board of seven (7) Directors, which shall exercise all the powers of the Cooperative except such as are by law, the Articles of Incorporation of the Cooperative or these By-Laws conferred upon or reserved to the members."—Section 1, Article IV, By-Laws

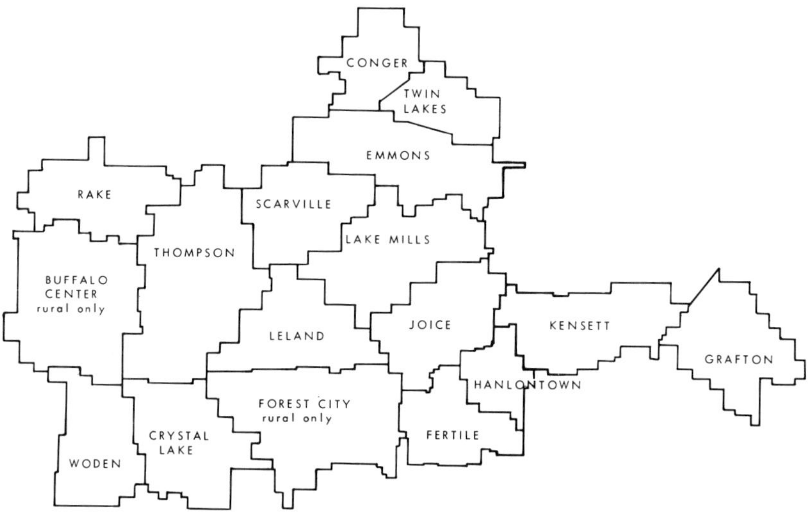
CONGER
TWIN LAKES
EMMONS
RAKE
SCARVILLE
THOMPSON
LAKE MILLS
BUFFALO CENTER rural only
LELAND
JOICE
KENSETT
GRAFTON
HANLONTOWN
FOREST CITY rural only
CRYSTAL LAKE
WODEN
FERTILE

Invention

"Mr. Watson, come here; I want you."

The words were spoken in an urgent and peremptory tone by Alexander Graham Bell to his assistant, Watson, who was in an adjacent room. Bell had just spilled a mixture of water and sulfuric acid on his work table.

Watson came running, not in answer to Bell's command, but to proclaim in an excited voice, "I heard you!" Watson had heard Bell's voice come over the experimental phone connection between the two rooms.

Bell had succeeded in sending the human voice over a wire—the Bell telephone was born! The date was March 10, 1876.

This first phone of Bell's was an outcome of his earlier discovery of a striking relationship between a thin iron diaphragm and a magnet wound with thin insulated wire. When the diaphragm was brought very close to, but not touching, the magnet, a current in the magnet caused the diaphragm to oscillate audibly in timing with the pulses in the current.

From this, Bell rationalized as follows: "If I could make a current of electricity vary in intensity precisely as the air varies in density during the production of sound, I should be able to transmit speech telegraphically."

This device into which he spoke the historic sentence to Watson had a voice-actuated diaphragm that varied the depth of a rod immersed in water mixed with sulfuric acid—perhaps the same liquid he had spilled—thus varying the electrical resistance and hence the current.

Other minds quickly utilized Bell's fundamental discovery to further improve telephone transmission. In 1877, Thomas A. Edison encased finely ground carbon granules, against which was pressed a diaphragm that responded to frequency modulations of the spoken voice by alternately compressing and releasing the carbon grains in the transmitter.

A battery powered circuit between the diaphragm and a receiver connected the speaker and listener, and vice versa. Magneto (generator) powered bell ringers were already in existence and were incorporated into the phone.

This practical application of the newly invented and quickly refined telephone spread like wildfire world-wide. Indeed, as early as 1878 there was a telephone connection between two of our Minnesota neighbors, the towns of Kasson and Mantorville, not too far from Winnebago County. On July 11, 1878, the Mantorville newspaper, the *Dodge County Republican,* printed, "The telephone is quite a success. Conversation is easily and freely indulged in between parties in Mantorville and Kasson. Music played in one office fills the other office three miles away with its melody, clear and distinct. Now, if it is only a success as a business venture, so as to make it a permanent institution, it will be of absolute utility to both places."

The above newspaper notice was suggestive of the prediction made that same year by the marvelously prescient Mr. Bell:

> It is conceivable that cables of telephone wires could be laid underground, or suspended overhead, communicating by branch wires with private dwellings, country houses, shops, factories, etc., etc., uniting them through the main cable with a central office where wires could be connected as desired, establishing direct communication between any two places in the city. Such a plan as this, though impracticable at the present moment, will, I firmly believe, be the outcome of the introduction of the telephone to the public. Not only so, but I believe, in the future, wires will unite the head office of the telephone company in different cities, and a man in one part of the country may communicate word of mouth with another in a distant place. I am aware that such ideas may appear to you Utopian . . . believing, however, as I do that such a scheme will be the ultimate result of the telephone to the public. I will impress upon you all the advisability of keeping this end in view, that all present arrangements of the telephone may be eventually realized in this grand system.

In fact, industry responded so quickly to the public's appetite for telephones that the Mantorville *Dodge County Republican* was reporting in its issue of July 18, 1878, that "the Messrs. Ware Brothers have returned the 'Bell' telephones, with which they have been experimenting and ordered those of the Edison patents to supply their places."

The first telephone connections ran a single wire between two places. When the places were some distance apart, it was practicable to run leads from the wire to additional phones in between, thus making a multi-party phone line.

The telephones were battery/magneto, i.e., a telephone had two 1-1/2 volt dry-cell batteries to provide voice power, and a hand-cranked magneto (generator) to provide ringing power. When the magneto was rapidly cranked, every phone on the line rang, so it was necessary to assign a ring code to each phone on the line. Mrs. Smith's phone code might be two short rings and one long, so when Mrs. Jones called her, assign a ring code to each phone on the line. Mrs. Smith's phone code might be two short rings and one long, so when Mrs. Jones called her, Mrs. Jones would make two short cranks and one long on her magneto. Mrs. Smith would know the call was for her; the others on the party line would know the call was not for them (although many a curious soul picked up the phone anyway for "listening in"—we won't use the pejorative "eavesdrop").

Naturally as more people came on the lines, they wanted to talk to people on the other party lines. And it quickly became apparent that a central switching center was needed to interconnect the various party lines in an area.

Area's First Switch

As early as 1896, in the home of Dr. G. M. Lee in Thompson, Iowa, there was a small switching device, tended by Mrs. Lee. Night and day,

The late Dr. G. M. Lee, Thompson, points to the first switchboard in Thompson, installed in his home in 1896. The circular board had several switch buttons, enabling connections from different directions, ranging from north to east to south to west. The board had a central rotating connecting switch. This primitive switchboard enabled the doctor to call back to his wife to see if he had any other calls in the area, and for her to leave word at a farm that someone had called in for him. Also, with the board, Mrs. Lee became Thompson's first telephone operator, switching farm and town calls.

seldom learned he was needed in another area until he arrived home and was told by his wife that so and so had ridden into town seeking the doctor. To help correct this problem, Dr. Lee and some of the farmers first put up four telephone lines, one each from the north, south, east and west coming into Thompson and joined by a switch in the Lee home. The switch was a crude affair, made of wood and about eight inches in diameter with metal buttons around the circumference for each line, and one button and a lever in the center with which to make contact with the calling line. When Mrs. Lee heard a ring she could either take a message or switch the call to one of the other three lines, which were party lines. The call might be a medical call, or a request from one of the farms on the line to be connected with a farm on another line. Or it could be a call from Dr. Lee asking if there were any messages before he started home. Also, Mrs Lee could call out to the farm where she knew her husband was, or to a farm in that area.

The Early Years

Later, groups of farmers called neighborhood meetings and decided to form little companies of their own and build a telephone line to tie in with a town switchboard. Towns and cities already had switchboards serving a great number of their citizens, the higher density of population making phone service practical and economical.

There were many little phone companies. Some would organize under the laws of Iowa as mutual stock companies, or mutual non-profit systems, or, in many cases, as cooperatives. Sometimes a group of 10 or 15 farmers would ante up enough money to build the line to town, and buy their own telephones and not be incorporated. In fact, except for a small minority of cases (Emmons and Conger in the RTA area), each farm family owned its own phone. Subscription rates ranged from $6.00 to $30.00 per year.

The great proliferation of such entities is reflected in the fact that today's Winnebago Rural Telephone Association consists of the amalgamation of sixty eight (68) area telephone systems and a cable TV system.

Forest City rural, for example, had 17 different telephone entities, either mutuals, unincorporated associations, or cooperatives. These had banded together to become the Farmers Switchboard Company, which bought and staffed a switchboard, which they installed in the same Forest City building that housed the town's commercial switchboard. There was a connecting cable between the switchboards; farmers would call their own operator, who in turn would call the commercial company operator who would connect the farmer with a local business or town resident, and vice versa.

In Thompson, there were actually 12 different systems that owned the joint switchboard in town, there being no commercial system serving Thompson.

The switchboards of all the small companies were located in towns of their areas. The earliest switchboards were magneto/battery equipped, the magneto for providing the ringing power, the battery for voice power. When Mrs. Jones wanted to call Mrs. Brown on a line other than her own party line, she rang the switchboard operator, who would switch Mrs. Jones' line to Mrs. Brown's and make the required magneto cranks for Mrs. Brown's special ring. Later a battery powered ringer replaced the magneto, giving relief to the operator's arm.

One can imagine the communication environment of these early, primitive, telephone systems before the Winnebago RTA came into the picture. With many "listeners in"—the greater the number, the weaker the transmission—with always a good percentage of the individual phones having weakened batteries, with the ravages of time putting many systems in various stages of disrepair. The advent of the power lines causing a hum on the telephone lines (see page 7), yes, indeed, those were the days of "whoop and holler" telephones.

The Party-Line Operators

Inconvenient? Yes, but the inconvenience was leavened by the role of the switchboard operator. The telephone systems were the newspaper gossip of the areas, and the switchboard operator was the newspaper's news, advertising and social departments combined.

The operator kept tab of everyone's comings and goings and whom they were likely to visit. If you called Mrs. Jones, the operator would know she wasn't home because it was Tuesday afternoon and she was with the coffee group at the neighbors, and would you like the operator to call her there. You would, and she did, and Mrs. Jones would be there.

The veterinarian would tell the operator where he would be on his next service call, so if he had a call, the operator would relay the message to him at the farm where he was working. In short, the operators served a grand duty in the community.

The operator was the exchange's secretary as well as its news-gathering organization. Sometimes she'd send out four long rings on the line, which on most exchanges signalled a general call. It might be the local store advertising the price of eggs on Saturday, or the livestock buyer giving that day's prices for hogs or cattle. Or, the grocery store would have a special—it would pay nine cents a dozen for eggs and sugar was on sale for one cent a pound. Each such announcement cost the advertiser fifty cents, which often was paid directly to the operator as a bonus.

The operators generally were housed in a small building or apartment, having a bedroom, or maybe two, with a couple of beds, small kitchen and bath. There was a sitting room where the operators worked.

An operator had a relief assistant; one or the other was required to live in night and day. The office usually closed at 10:00 in the evening; some offices had surcharges of twenty-five cents or fifty cents for any calls from 10:00 until 6:00 a.m.

Operator wages were low, maybe $35, $50 or $60 per month, and that wasn't for a 40-hour week. After the Winnebago RTA took over, the operators were put on a 40-hour week with time and one half for overtime at a rate of pay based upon the government's minimum wage. The RTA costs went up dramatically, far in excess of the old companies', which were not subject to the wage and hour law.

Early Technology

The early systems were economical "single wire and ground return" circuits, the equipment connecting with a ground rod at the receiving end, the ground serving to complete the circuit back to the switchboard. More than one wire, of course, led out from the switchboard; there was a wire for each party line. Later on, a second wire was added, dubbed a metallic return, to complete the circuit instead of the ground return. The change was accelerated by the advent of the power lines. The power lines produced an electrically induced interference; the 60-cycle alternating current in the power line created a magnetic flux surrounding the line and this flux radiated to the adjacent telephone lines on the other side of the road. (Power lines ran on one side of the right-of-way, telephone lines on the other, the line which was first having the choice of sides.) The flux induced an alternating current in the telephone line that produced a 60 cycle hum—sometimes almost a roar—in the listener's receiver/ear.*

The problem was substantially solved by adding a second wire to the pole instead of using ground return. This cured most of the interference, but not all. One of the two parallel lines, being closer to the power line, picked up some interference. To further reduce the interference, the two phone lines were criss-crossed every quarter mile, the alternate interferences tending to cancel one another.

The line noises, the faint speech transmission because of weak batteries, the party-line interference, etc., etc., were all eventually eliminated by replacing the overhead line with underground cable. But before the Winnebago RTA improvements, hardly any farmer used the line for

"Sometime before 1949, Tommy Adams, Camp Point, Illinois, had invented a system of superimposing high frequency radio waves over REA power lines, with the resulting radio waves being picked up by a radio receiver in the lineman's truck, up to the FCC allowed maximum distance of 1,000 feet from the power line. The February, 1949, issue of the Winnebago REA house organ, *Electrik*, informed its readers that it could not "advise them concerning the feasibility of rural telephone by the use of radio carrier waves over the same power lines through which electricity passes." The REA management no doubt by then was anticipating a more practicable telephone installation, because the same article advised the reader that "there is now a bill in Congress providing for a rural telephone loans act similar to the REA act. This bill will make loan funds available to establish rural telephone Co-ops., similar to the REA systems. We believe that there is a possibility this act may be passed by Congress during this session. If you are interested, it is advisable to write a letter to your congressman or senator relative to your wishes."*

business because he didn't want everyone on the line "listening in." The telephone was simply a social and gossip communicator, not the great business tool it is today. In fact, many of the men in the early days would not talk on the telephone. Usually, if a man wanted to talk to Joe Smith, he would have his wife place the call and give his message to Smith's wife for relaying to Smith. You might say the farmer was "mike shy," and with good reason.

Mrs. Andrew Rasmusson is shown smiling at the old switchboard in Emmons, which would be replaced by the RTA's automatic dial exchange. Mrs. Rasmusson was a switchboard operator for the Farmers Telephone Company for 26 years. Looking on are Mrs. J. A. Wilson, switchboard operator, and S. R. Emerson and Jim Beaudy, manager and maintenance man, respectively, of the Farmers Company.

The Beginning of the RTA

The Winnebago RTA had its genesis in the late 1940's, the outcome of an idea in the mind of Glenn Bergland, who, at the time, was chairman of the Communications Committee of the NRECA—the National Rural Electric Cooperative Association, comprised of a thousand-plus rural electric cooperatives from across the country.

A continuing subject of discussion at the NRECA's annual meetings, particularly by the members of the Communications Committee, was the need to improve communications between co-op headquarters and their men in the field and their customers.

The cooperatives had begun to use two-way radio communication with their maintenance men. The Communications Committee addressed the task of coordinating the various FCC radio frequencies used throughout the nation and the solving of other intersystem problems.

In addition, the committee faced the challenge of the inadequacy for reporting customers' electrical outages via the "whoop and holler" rural phone systems, which were subject both to the hazards of physical deterioration and the possible absence of operators in attendance at all hours of the night.

"How wonderful it would be," Glenn's committee said, "if all of rural America could have modern dial telephone service." The reporting of electrical problems would be greatly facilitated and a corollary result would be a great benefit to the farmers of a modernized telephone system (and in particular, Chairman Bergland mentally reflected, to the farms served by the Winnebago REA).

To build and equip a modern rural telephone system nation-wide, however, would require huge sums of money. The committee decided the answer lay in having a federal lending program similar to the REA Act. The committee proposed to the NRECA resolutions committee on two successive years that it adopt a resolution empowering and authorizing its executive secretary to draft an amendment to the REA Act to expand REA's loan program to include rural telephone systems.

The second year the resolutions committee did accept the communication committee's resolution. It was presented to the NRECA membership, who voted that the NRECA executive secretary, Clyde Ellis, help draft an amendment to the REA Act, similar to the Poage-Hill bill, to provide for telephone loans.

The powerful lobby of NRECA prevailed upon the Congress to consider the bill; Representative Poage of Texas, the chief floor leader and advocate of the bill, pushed it through Congress. It was signed by President Truman in November, 1949.

The REA telephone loan amendment, which made possible the orderly and rapid transition of America's rural telephone systems from backward to modern was an example of Glenn's efforts in behalf of not just the farmers in Winnebago County, but farmers nation-wide. Later in this book, the reader will learn of other achievements by Glenn in behalf of telephone cooperatives in addition to Winnebago's.

Board members of the Winnebago Rural Electric Cooperative, 1940: From the left, C. M. Palmer, M. J. Rygh, A. D. Hartman, Elmer Bergland, A. T. Grosland, Archie Anderson, R. E. Aukes.

Winnebago REC

The Winnebago Rural Electric Cooperative Association (Winnebago REC, formerly called the Winnebago REA) was incorporated in 1938 for the purpose of providing an electric distribution system for farms in the Winnebago county area.

In 1936, Congress had passed the Rural Electrification Act to make available low interest, long-term loans for the electrification of the nation's farms, and on February 9, the Winnebago REA received approval of its first REA loan—$288,000.00.

The Winnebago board of directors, with approval of the federal REA, hired Glenn Bergland as project superintendent, starting April 1, 1939.

Glenn was given the responsibility for the construction, operation and management of the Winnebago REA system.

Glenn's Background

At the time of his employment by REA, Glenn was a student at Luther College, Decorah, Iowa. After graduating from high school, Glenn worked for four years before entering Luther. In his final year and a half at Luther, he worked part-time as a student assistant in the physics laboratory, for which he received credits on his room, board and tuition. He had completed his regular courses in 3-1/2 years, and this let him accept REA's April 1 employment offer. Glenn graduated from Luther with majors in mathematics, physics and chemistry. Later he became a member of IEEE, a national Engineering Society.

At age 15, working for his father, who sold Delco generating systems, Glenn made his first installation of the Delco system and wiring (with the help of the farmer) on a farm southwest of Lake Mills. Glenn started his new job at REA as a 'greenhorn', but with a credible technical background from his work and college training.

The new job presented a formidable challenge. There was not a pole in place, not a wire bought or strung, nor easements for right-of-ways, nor were there transformers, meters, insulators, or any of the myriad items required for an electric system. Employees, trained, or to be trained, were yet to be hired, nor were there buildings or offices for the new project. Nor was there a source of power; in the summer of 1939, however, Glenn negotiated a wholesale power contract with the city of Lake Mills.

Folks, this was 1939. Less than one American farm in 20 had electricity. Oh, there were a few farms in Winnebago county that tapped off the Interstate Power Company lines which passed their place, and some of the more affluent farmers had Delco electric generating systems, but essentially, it was sunlight and kerosene lamps, horses and biceps, and a few tractors . . . in April, 1939.

Less than a year later, as proof of the can-do performance of the young REA manager, the Jake Anderson farm south of Lake Mills was energized, Winnebago county's first REA hookup.

From the January 3, 1940, *Lake Mills Graphic* newspaper: "Long-time hopes and plans for rural electrification of Winnebago county farms became a reality this morning when Glenn Bergland, project superintendent, threw the switch at the Lake Mills municipal light plant, which energized the first REA lines throughout the county."

Item: Illustrative of Glenn's axiom, "Do it right and do it to last," was his response to the REA ultimatum that more than 40 percent of the transformers on the project be 1-1/2 KVA. These small transformers would not long have sufficient capacity to handle the average Iowa farmer's electric requirements. The Stanley Engineering Company was the engineer on the project and Glenn went through their staking sheets and changed most of the 1-1/2 KVA's to 3 KVA and 5 KVA. When REA headquarters in Washington received the final construction inventory, they refused to grant the additional loan requisition for the more expensive higher voltage transformers. After several discussions with Glenn, they sent an engineer from Washington to evaluate the situation. He checked a number of the farms and told Glenn he would approve the changes, but "don't do it again." In the continuing construction projects, however, Glenn proceeded to make the changes, with similar objections from the government, but with a subsequent okay of the changes. The government wanted cheap construction so as to build more electric lines in the beginning and did not take into consideration the future costs and requirements. As a result, in a few years many other REA's had 1-1/2 KVA transformers in the junk pile. The Winnebago REA let contracts for its first three construction projects, but from then on, did its own engineering and construction. As a result, the later lines were built with heavier wire, stronger poles and shorter spans between poles. These lines withstood the heavy weight of most ice storms, in sharp contrast to those built by the contractors.

REA Telephone Lending Act

For at least two years before the passage of the REA telephone lending act, Winnebago REA manager Glenn Bergland discussed rural telephone service with the Winnebago REA board. On many occasions, Glenn encouraged the board to consider by what means they might be able to bring better telephone service to the area.

This would supplement the REA's two-way radios and greatly facilitate the reporting by farmers of electric outages, besides providing better phone service.

On October 28, 1949, the Rural Telephone Lending Act was signed into law. It provided that twenty-five million dollars be made available for telephone loans, under the jurisdiction of the federal REA.

"Maybe we should take a leadership position in our area," Glenn told the board, "and help the mutuals to organize into one viable cooperative so as to be large enough to be self-supporting, with sufficient field staff, office and personnel to provide proper maintenance."

Glenn questioned, "Shall we invite the mutuals in for an information meeting and see if they want to join together with our help in organizing?"

Some of the board members doubted whether a telephone cooperative could be as successful as the electric cooperative, but authorized Glenn in November, 1949, to invite the mutuals to a meeting to discuss consolidation.

Consequently, fifty-seven mutual phone companies in the following communities were invited to send representatives to a meeting November 26, 1949, at the theater in Thompson: Rake, Titonka, Woden, Crystal Lake, Thompson, Scarville, Leland, Emmons, Joice, Hanlontown, Fertile, Buffalo Center rural, Forest City rural, and Lake Mills rural. (Phone systems within the three last mentioned towns proper were operated by independent companies.)

Manager Bergland, gave reasons supporting a consolidation of small mutual companies:

First of all, Glenn said, loan funds in the new REA Lending Act would go to the larger, viable operating units that are financially strong enough to repay the loans.

Secondly, most small phone companies are confronted with a difficult labor problem—they are too small to hire a full time maintenance man, so it usually becomes the duty of a farmer to act as lineman on a part-time basis using time he can ill afford to give up, especially during his busy season; and he is asked to collect dues, all of which is usually an accommodation on his part, at little or no remuneration. Even when full-time maintenance men are employed, they often are not utilized to their fullest capacity due to the lack of supervisory assistance in planning and engineering their work.

"There would be an organization adequate to handle maintenance, repairs, engineering, accounting, bookkeeping and collecting," he said.

The small companies lack the funds to finance rebuilding of their systems, many of which have lines and switchboards that are thirty years old, Glenn said.

"If the existing mutuals merge, it could be possible to change all phones to dial, to install automatic dial switchboards serving each community, with each switchboard connected to a remote toll center with answering operators," he added.

One hundred seven mutual representatives attended the meeting. Also in attendance were personnel from the various larger independent telephone companies of Iowa. They posed questions doubting the ability of a cooperative to succeed.

"The independents wanted to 'swallow up' the small mutual companies," Glenn says today.

Attorney Dudley Weible of Forest City discussed the legal aspects in effecting a consolidation.

The meeting adjourned with the small companies agreeing to poll their members as to their interest in better telephone service. Questionnaires were given to the mutuals to use in sounding out their members. Included were questions asking whether or not the existing systems were mutuals, whether or not incorporated, how many members were there, how many miles of line, whether single-wire-to-ground or metallic return, condition of the lines, current dues charged, whether in favor of a unified system, and what they would pay for good dial telephone service.

Those in attendance expressed their opinions, requesting the electric Co-op to get the ball rolling.

Favorable Response

At a second meeting on December 12, 1949, it was reported that in the communities of Scarville, Thompson, Buffalo Center rural, Joice, Lake Mills rural and Forest City rural, more than seventy-five percent of the telephone subscribers were interested in a merger.

At this meeting, manager Bergland discussed the idea of putting all assets of all companies into one pool. The REA Co-op would, in turn, have an appraisal made and render to each individual a personal capital credit for his proportionate share of his old company. The credit would be paid to the subscriber at a later date when it was financially feasible for the new company to do so. It was suggested that the new cooperative would assess each member a $10 refundable membership fee.

A motion was made requesting and authorizing the Winnebago Rural Electric Co-op Association to do the organizational work and operate the new system, to be called the Winnebago Cooperative Telephone Association. The Electric was to formulate merger plans, make evaluations, and draw up Articles of Incorporation and By-laws.

Sitting in the audience was Harold Hove, who farmed eight miles north of Thompson. Hove wrote on a card, "I so move," and passed the card to Emory Joe Rognehaugh. Rognehaugh wrote on the card, "I second," and handed the card to Manager Bergland. Manager Bergland then turned to the audience (members) and said, "It has been moved and seconded that we take the initiative to organize the old telephone companies into one telephone cooperative and operate it as a separate cooperative—that is, if the companies so desire."

President Elmer O. Bergland called for a vote. "All in favor say aye; opposed, the same. The aye's have it. The Board will make a decision if it is feasible to assume the extra responsibility."

After this December 12 meeting of the area mutuals, the REA Board of Directors authorized attorney Weible, in consultation with manager Bergland, to draw up tentative Articles and By-laws for a new Winnebago Cooperative Telephone Association.

1950

On January 12, the REA board, Manager Bergland and Attorney Weible again met with the delegates and reviewed with them the newly drafted merger agreements, articles and by-laws, and proposed rates. Also in attendance were other interested subscribers.

A rate study had been made by a retired Bell Systems engineer, Vincent Lobaugh, and it was thought possible to charge $3.75 per month for eight to ten party rural service.

The delegates voted to accept the proposals and authorized the REA board to proceed with the organization. Arrangements were made for Manager Bergland to attend individual meetings of twenty-four prospective companies relative to consolidation.

On January 15, 1950, the seven board members of the Winnebago REA met, both as individuals and as charter organizers to organize the telephone cooperative. The seven approved the tentative Articles and By-laws and moved their adoption, and directed that they be filed with the Secretary of the State of Iowa.

Two Co-ops, One Board, One Manager

To maintain a workable arrangement between the two cooperatives, it was necessary for the seven to elect themselves to be the original Board of Directors of the telephone cooperative, with the same officers. They were: Elmer O. Bergland, Thompson, IA., President; M. J. Madson, Emmons, MN., Vice President; H. D. Hartman, Buffalo Center, Secretary-Treasurer; R. E. Aukes, Buffalo Center; M. J. Rygh, Lake Mills; Archie Anderson, Buffalo Center; and Gustaf Gustafson, Forest City. (Glenn and Elmer Bergland are not related.)

Immediately following the January 15 meeting, a discussion ensued as to how the telephone Co-op would be managed. It was agreed that the employees could not have two bosses, nor should the manager work for two different boards.

The Electric was a viable co-op with resources and manpower. It was decided to enter into a management contract between the two co-ops whereby the Electric would manage and operate the telephone co-op on an actual cost basis.

Glenn Bergland was named manager of the telephone cooperative. He would receive no extra compensation for his work with the new telephone cooperative, which would absorb about ninety percent of his increased workload during the next two years.

"To me it was a personal challenge to effect the modernization of rural telephony," Glenn says.

Attorney Weible and manager Bergland were instructed to draft a proposed agreement. The detailed agreement allocated the manager's time, office costs, heat, light, salaries, trucks, etc., between the two co-op's.

At the next board meeting, the proposed joint management and operations agreement was approved by the identical Telephone and Electric boards. Thereafter, board meetings of the two cooperatives would always be conducted separately but usually on the same date.

$100 Per Subscriber? No!

The federal administrators of the 1949 REA Telephone Loan Act ruled that eligible telephone cooperatives must show as equity in their capital structure a $100.00 cash deposit from each member (in addition to their $10 refundable membership fee). The commercial telephone companies were exempted from this provision, the administration taking the position that they were more stable financially than the cooperatives.

Manager Glenn Bergland realized immediately that asking $100.00 from each member would be fatal to any consolidation—the members would simply refuse to pay. (A hundred dollars in 1950 equates to possibly a thousand dollars in 1990.) He kept the information and his fears to himself and his board of directors; he was determined to make every effort to have the provision rescinded, and meanwhile, he proceeded with the consolidation.

In any event, it was several months later before Glenn had finally convinced those in charge of the loan program to cancel this equity requirement.

Because this was a victory with far-reaching effects—a battle fought and won not just for the Winnebago RTA but for all the telephone cooperatives across America—we will have Glenn narrate in his own words the particulars of the drawn-out struggle:

By Glenn Bergland

In March, 1950, the National Rural Electric Cooperative Association of 1,000 electric systems had its annual meeting in Chicago. A guest speaker at the meeting was Claude R. Wickard, REA Administrator, who was also in charge of the REA telephone loan program.

I requested a meeting with Mr. Wickard and was given an appointment for the following evening, for just 15 minutes; scores of persons from throughout the country were demanding Mr. Wickard's time, to learn how to utilize the loans program to accelerate the rebuilding of rural telephone service in their areas. In attendance at my meeting with Mr. Wickard were Dick Dell, loans application office; Ray Lynn, engineering division chief, and representatives from the appraisal and legal divisions.

I argued that we had sufficient equity in behalf of our subscribers because the old lines not only had some current value, but also a great potential value if incorporated into a new, modern system.

However, Wickard and Associates knew the lines were worthless and, really, so did I, but I used it for an argument. Also, it could be stated on our books that our right to do business had a real value.

That 15-minute appointment turned out to be a two-hour meeting while the others outside waited their turn. I was arguing against all the staff until, finally, about half-way through the meeting, Mr. Wickard began coming to my defense. He told the engineering and finance chiefs that they should set a date with me as to when they'd be out to Winnebago to see what we had.

"Here is a fellow," Mr. Wickard said to his associates, "with 27 effective mergers and last night a group of my own Indiana people from seven different telephone companies each wanted an individual loan. I asked them to join together, but they threw their backs to the wall and said such was impossible and couldn't be done. Here we have someone with a consolidation of twenty-seven and you gentlemen will not even look at him. Go out there and see what he has and see what we can do on the capital credit requirement."

Dick Dell and five others came to Winnebago the latter part of March. Included were George Haggard, deputy administrator; Art Gerth, chief of the loans division; Ray Lynn, chief engineer; and a representative from the legal division.

We drove the lines and I tried to point out their value. My guests replied that although the lines had no value, they would reduce to $50.00 from $100.00 the equity requirement.

I argued strenuously and finally about noon Dell said he had called the REA administrator in Washington. Dell said to me, "We have good news for you. We'll cut it down to $35.00." We had a stack of consolidation papers about two feet high on the table. I walked up to the table and patted the stack and said, "We'll burn these first." Then I walked out of the room.

Dell followed me into the other office and said, "Glenn, you can't do that." I said, "Yes, I can. The people will be very happy, I am sure; our president sitting out there and our secretary are also in favor of discontinuing our application due to your exorbitant request."

Dell asked, "What will you tell the people?" I said, "It's easy. We'll just tell them that due to your exorbitant demands we saw fit to rescind our loan application, and we'll send out publications and public notices to all of the major national telephone magazines in the country, and also to *Readers' Digest* and to *Telephony*, the Telephone Engineer and Management magazine, all of which have up to now been very critical of your program and have at many times printed adverse comments. I'm sure this will be more ammunition for them."

Dell and his associates exclaimed, "Glenn you can't do that." I replied, "The monkey rides on your back now, it's not on mine anymore," and I left the room.

The group had planned to leave the next morning and so that evening we had dinner together. They started to bring up the subject, but I said, "Forget it—it's history. I am not in the telephone business anymore, let's enjoy our steaks."

Finally, they said, "Let's ride the lines once more tomorrow." So we did, and as a result they urged me to keep working. I replied that we didn't want to pour any more money into the project unless we could be assured that we'd get a loan on our conditions.

I did tell them we'd appraise the lines for their replacement cost, less accrued depreciation, to determine what equity we had, if any. We hired Vincent Lobaugh, a retired Bell engineer for the appraisal. He estimated a replacement cost, less accrued depreciation, of $53,000 (a fictitious value) for the nine exchanges in our first rebuilding project, and we forwarded this figure to Washington. (We later used it as a capital credit for allocation back to the members.)

Sometime in May Dell called and said, "Glenn, we have wonderful news for you. In addition to your refundable membership fee of ten dollars, an additional five dollars cash equity payment will make your project feasible."

To me, the additional five dollars was laughable. "Dick, do you mean five measly bucks makes this project feasible or not feasible?"

He answered, "That's what we came up with." I said, "Forget it; cancel the loan. We don't want any part of an endeavor that close to being a success or failure." The next day Washington called to say they were approving the loan with no equity payment required.

Early Financing

During the next two or three years, the telephone cooperative became indebted by more than $100,000.00 to the electric cooperative.

Without the resources of the electric cooperative, it would have been difficult for the telephone cooperative to succeed. It now had to comply with the provisions of the federal wage and hour law—time and one-half for overtime and a minimum wage for the switchboard operators who had previously received low monthly wages for up to 24 hours per day on the job. There were, of course, construction, payroll, equipment, etc., expenses.

The increased expenses of the new Co-op were far in excess of the revenues derived from the old magneto companies. As a result, manager Bergland, who had been making revenue and expense projections for the new system and was greatly concerned about the viability of the new cooperative, went to Washington, D. C., to review his projections with the REA finance and loans division. He was encouraged by their telling him that "because of the fact that you are worried, we are sure your new telephone company will succeed."

Opposition

The Winnebago RTA was not organized without its adversaries. Although 95 percent of the members and directors of the area's mutuals favored consolidation, many working hard to help achieve it, a few were against it. Perhaps some thought their lines were undervalued and could be more profitable if independent; bear in mind, too, that of the more than 200 directors in the small mutuals there could be some who were envious of their authority and didn't want to relinquish their directorships.

Harmful rumors were floated, insignificant, but, absent any detail, they were vicious. "I spent many Saturdays, Sundays and evenings calling on people who needed an explanation of a rumor. I'd give them the truth to squelch the rumor, and it all took a considerable amount of time and effort; I was harassed to the point that life became a little difficult." Manager Bergland said.

In Crystal Lake and Kensett, the Cooperative was threatened with lawsuits. These were nuisance suits by just a few members; in fact, their effect was to unite the other members in the RTA's favor. Attorney Weible did outstanding work in showing the would-be suitors that they had no grounds for suit.

Central Iowa Telephone (owned by Sutherland Dows with offices in Cedar Rapids) owned the commercial systems in Forest City and Buffalo Center, but nineteen small mutuals owned the Forest City rural lines (and Leland), with their own switchboard in Forest City, called the Farmers Switchboard Company.

Upon learning that Winnebago RTA was organizing the seventeen rural mutuals, Central Iowa began building a few lines north, south, east and west of Forest City, which interspersed with the mutuals' lines. They then claimed that the Cooperative was duplicating an existing service. Central Iowa made a request to REA in Washington that the Cooperative's loan application be rejected.

A hearing on the matter was held in Washington. The day before the hearing, manager Bergland, attorney Weible and RTA board president Elmer Bergland gave Mr. Wickard a complete orientation, including maps, of the Forest City rural area and Leland, and so on the day of the hearing Mr. Wickard had complete knowledge of the layout of the lines. He saw Central Iowa's spurious claim for what it was and ruled in Winnebago's favor.

Consolidation Procedure

To bring a mutual into the Cooperative, three separate meetings with the mutual were required.

The first meeting was with the mutual's board. The Winnebago RTA was further explained and questions answered. The election procedure was explained. The board would then call an election meeting of its stockholders.

At the second, election, meeting, the merger was explained to the mutual's stockholders and ballots were distributed. Mail ballots and proxy forms had previously been provided.

A two-thirds affirmative vote was required for the merger.

The third meeting was with the mutual's board to adopt the merger plan and transfer the mutual's assets to the Cooperative. There were deeds to be assigned, entries to be made in abstracts, rights-of-way to be given, etc. Attorney Weible was an excellent legal supervisor.

Imagine the magnitude of the task! In the months of January, February and March, 1950, 43 mutuals were brought into the Cooperative. At times there would be two meetings in the morning, two in the afternoon and two in the evening—almost a fixed schedule of 10:00 A.M., 1:00 P.M., and 7:00 P.M.

Representing the RTA at the meetings were Glenn Bergland and attorneys Weible and Harry Slife (Dudley Weible's partner), and Charles Peckham, a retired inspector for Iowa's Department of Agriculture. Mr. Peckham was employed by the Cooperative to solicit easements for right-of-ways.

It was clear that once the mutuals started to merge, the process had to steadily continue; a delay of any length would jeopardize the entire project.

"We had started the ball rolling and we had to keep it rolling," manager Bergland said.

The first 10 companies had joined by the 25th of January, 1950. They and their officers and board members were:

Forest City rural: Citizens' Mutual, Sam Honsey, Chairman; Peter Osheim, Secretary; Gilbert Samuelson, Richard Haugland, Oscar Jordahl, Norman Brackey, H. O. Lovik

Otis Mutual: Dale Jacobson, President; Terry I. Thompson, Secretary; Eugene Otis, Merle Otis, Russell Lewis.

Forest Township Mutual: Gordon Kloppen, President; Ralph Nelson, Secretary; Oscar Ness, O. J. Charlson, J. H. Goranson.

Rake: East Norway, Ludvig Jacobson, President; Carl Havnen, Secretary; Nordahl Skogen, Elmer Tollagson, George Jergenson.

Thompson: East Side, Odin Davidson, President; Art Hauan, Secretary; Nels Ellingson, Conrad Reines, Alfred Swenson, Walter Anderson, Bernard Arnevik.

King Linden: Dan E. Lemme, President; Orland Anderson, Secretary; Claire Doyle, Lippo Asmus, Leonard Sandell.

Lake Mills rural: Silver Lake-Bristol, Lewis Horvei, President; Tosten Groe, Secretary; Carl Christianson, Roy Bergan, John Bjelland.

Newton Mutual, Elmer Bergland, President; Iver Solomonson, Secretary; Andrew Solomonson, Clarence Olson, Art Carlson, Milton Charlson.

Leland: Leland Mutual, T. O. Carlson, President; F. W. Bowman, Secretary; Otto Nolte, S. B. Nelson, Francis Dunbar, Ed Ambroson.

Scarville: Scarville Farmer's, Emil Hermanson, President; Milford Wogen, Secretary; Ben Nelson, treasurer; John Solomonson, Manager; Selmer Cleven, George A. Larson.

And the ball kept rolling, like a growing snowball. By the end of March, 1950, more than 100 organizational meetings had been held and 43 mutuals had been merged into the RTA.

The 43 were served by the following nine switching centers: **Lake Mills rural; Scarville,** town and rural; **Leland** town and rural; **Forest City,** rural; **Crystal Lake,** town and rural; **Thompson,** town and rural; **Buffalo Center,** rural; **Rake,** town and rural, and **Joice,** town and rural.

These nine exchanges comprised the first construction project. They were followed by Emmons, Conger and Twin Lakes, in Minnesota, in 1953; Grafton, early 1960; Fertile, summer, 1960; Kensett, 1961; Woden, 1962, Hanlontown, 1967, and Lake Mills town, 1982.

At the time Grafton requested membership in the RTA, Grafton was not contiguous to the RTA area. The RTA board accepted Grafton notwithstanding, and with the entry of Kensett in 1961, all the RTA exchanges were again contiguous.

The Minnesota Public Service Commission was reluctant to approve the transfer of Emmons/Conger/Twin Lakes because they believed cooperatives were not as service conscious as the independent operators in the private systems. But after the exchanges' members voted unanimously to join the RTA, the Commission gave its approval of the transfer and also of the rates established by the RTA for the three exchanges.

Farmers Telephone Company of Emmons and Conger becomes 45th independent telephone company to join the RTA when F.T.C. president Andrew Larson, center, and Bernie Loberg, secretary, sign consolidation documents. Witnessing the ceremony are S. R. Emerson, F.T.C. manager, left front, and, standing, Jim Beaudy, F.T.C. maintenance man, Malmin Madson, RTA vice president, and Manager Bergland. (1953)

Thompson Telephone Co-op

The following is both an interesting episode in the history of the pre-RTA Thompson mutual phone companies and an account of Glenn Bergland's first involvement in telephone.

The time was the middle 1940's. the Thompson exchange comprised several farm mutuals and a town company, and together they owned a switchboard and building in Thompson under the name of The Switchboard Company.

Over the years, the shares of the town company had fallen into the hands of a few individuals who, at the time of this story, had just sold the town company to Central Telephone Company of Iowa, for $5,000.00. Central in turn demanded that the farmers sell to Central their shares in the Switchboard Company, now more than 50 percent owned by Central

Telephone, which would then provide switching service to the farm mutuals at a monthly rate.

The farmers rejected Central's demand and appealed to the Thompson Business Men's Club for help. The Club was sympathetic to the farmers and appointed a committee to try and help them. Committee members were Glenn Bergland, chairman; Theodore Klemensrud, Sanford Overrlie, Otto Alquist and William Erickson.

The committee discovered that Central did not have a town franchise to use the streets and alleys, and so in effect were operating illegally.

Chairman Bergland proposed, and the committee approved, the organization of the Thompson Telephone Cooperative. The cooperative was chartered by the state, and committee member Klemensrud, who owned the *Thompson Courier*, published in that paper several articles about the new cooperative and its intention to seek a franchise through a city election.

Shortly after the articles appeared, a Central Iowa representative, Richard Thompson, visited Glenn. "What are your intentions?" he asked Glenn. Glenn answered that the new Thompson Telephone Cooperative planned to serve the town, but because it was short of funds, it would first put up cables in the alleys in the business block, and connect them to the business firms and collect their toll revenues through connections with a new Farmers Switching Center, which the farmers would build.

Thompson exclaimed, "You can't get away with this, you have no legal rights!"

Glenn replied that Central was using the streets and alleys without a franchise, and "we intend to get one by town vote and extend our lines as finances permit."

Thompson replied, "We will pull out our lines."

Glenn said, "You do not dare. You would be legally liable until we can service the whole town, at which time the council can order you out."

Following this, Central invited the farm companies to a meeting and the farmers asked the business club committee to attend, also.

Both sides became vociferous at the meeting, but Central's arguments were weakened by its not having a franchise. Finally, in frustration, Central said, "We'll sell the Thompson Telephone Company to you."

"How much?" asked a voice from the group.

"For $5,000, just what we paid for it," Central answered.

"We'll take it!" replied Otto Alquist, a local lumber dealer.

So Otto loaned the $5,000 to the new cooperative and it was in business.

Glenn Bergland was named president of the co-op, Klemensrud, Vice President, and Alquist, Treasurer.

Glenn was also elected to represent the new co-op in the switchboard company. He served as its secretary-treasurer until its merger with the Winnebago RTA.

"Those were the days," Glenn recalls, "but all the small mutuals lacked the funds necessary for them to give good service."

Switching Contracts

Central Iowa Telephone Company (which later became General Telephone Company) owned the Forest City and Buffalo Center town exchanges, and M. O. Godtland owned the Lake Mills town exchange. Consequently, it was necessary for the RTA to negotiate contracts for switching the farm lines through the switchboards of the three towns.

Sutherland Dows, president of Iowa Electric Power and Light Co., was the majority owner of Central Iowa Telephone. Dows was a fine gentlemen, Glenn says, but a tough negotiator.

The Winnebago RTA was a national testing ground as to whether or not the new telephone cooperatives would succeed in their attempt to operate on an equal basis with the commercial telephone companies. Those were momentous times. A great number of the large independent companies were adverse to the REA loans program and consequently Central Iowa leaned on them for expertise and help. Sutherland and his associates tried very hard to hinder the progress of the Winnebago RTA.

They also called on the United States Independent Telephone Association, USITA, to help them negotiate with the RTA.

"A Colonel Henry of Ohio, a USITA member, was on hand with ideas for Dows to use against us," Glenn recalls.

"Early in our negotiations, Co-op president Elmer Bergland and I jointly signed with Sutherland a letter of intent on the principal terms of a switching contract for Forest City and Buffalo Center; but when we tried to work out the contract with Dows' staff, they refused to comply with the terms of the letter," Glenn says.

"President Elmer Bergland and I then met personally with Dows. I reminded him of his signed agreement and asked him pointedly if the letter was no longer in effect."

"I'm sorry," Dows said, "but my staff says the letter's terms are unacceptable."

Elmer and I sat in stunned silence.

He said, "I'm sorry, fellows, it can't be helped."

"Elmer and I left Dows' office, but we returned to Cedar Rapids almost once every month during the winter of 1950-51, driving through weather at its worst, before we finally hammered out a switching contract. Although we threatened to build our own switchboards in Forest City and Buffalo Center, we learned that such would be very costly because of the high rates for switching from our boards to the town lines. Also, at that time, the Iowa Law was adverse to two switchboards in a town."

Glenn recalls an interesting sidelight of the meeting with Dows:

"In the hallway shortly after the meeting, Dows took me aside and asked, 'By the way, are you interested in changing positions and working for me?' I replied, 'No,' and that ended the conversation."

M.O. Godtland who owned the Lake Mills telephone exchange, had earlier taken the position he would wait until RTA settled the switching issue with Forest City and Buffalo Center, after which he would sign a switching contract with RTA.

In later years, the Winnebago RTA joined USITA, and Glenn served on several USITA committees.

First REA Loan

On June, 1950, the cooperative mailed a formal telephone loan application to the REA in Washington, asking for $452,000, to be used to bring better telephone service to 2,235 phone users and 305 prospects. The application cited anticipated annual revenues to the cooperative of $75,600, plus a portion of long distance charges, and listed as assets the $53,000 appraised value of the old lines and $23,350 in ten dollar membership fees.

With the application went maps showing the age and routing of 746 miles of telephone lines, and maps pinpointing the location of all telephone users and prospective users.

The Winnebago RTA was the first in Iowa and one of the first five in the nation to file a viable, constructive application for a telephone loan.

On October 27, 1950, the REA approved an allocation of $552,000 for a telephone loan to the Winnebago RTA. The amount represented an increase of $100,000 over the $452,000 in the loan application.

The Winnebago REA/RTA was the first joint operation or dual cooperative in the nation to get an REA telephone loan. It was followed later by a dozen or so joint operations across nation. Manager Bergland was first joint manager (REA/RTA) in the nation.

The loan approval followed a hearing only days earlier in Washington in which REA Administrator Claude Wickard overruled objections to the loan allocation filed by the Central Iowa Telephone Company.

The loan funds were to be used to rebuild 325 miles of existing telephone lines, to build 385 miles of new lines, to lay underground lead cables in the towns, to buy and install new dial switchboards, and to furnish subscribers with new dial phones.

The loan amount would not be forthcoming in a lump sum, but would come to Winnebago as payments once a month for work to be done in the following month as detailed in requisitions submitted by Winnebago and approved by the REA.

Loan terms provided for repayment in thirty-five years at two percent interest, with payments due every three months.

1942

Glenn Bergland, REC manager, with early office staff, 1942. From the left: Palmer Hamlin, Betty Alexander, Alice Christian, Manager Bergland.

1951

The January, 1951, *Electrik* advises farm members of the pre-RTA phone companies that any unpaid bills from the old companies must be paid to the new telephone cooperative and "even though you have stock in the old companies, it is necessary that a $10 membership be paid to the telephone cooperative.

"It is believed that because of the war (Korean War) there will be a scarcity of materials for, and resulting delays in, the rebuilding of the old lines."

The *Reader's Digest* published an article highly critical of the REA's telephone lending policies. The Digest had a reporter call on RTA manager Glenn Bergland, who was very complimentary of the REA's lending procedures. No article resulted from the reporter's visit.

On March 31, 1951, the Winnebago RTA received approved REA loan documents. Requisitions for loan funds could now be made.

On June 1, 1951, the new telephone cooperative began collecting dues and mailing telephone bills for service from March 1.

For the period during magneto (old) operation until dial conversion, rates were announced to be $2.00 per month for rural multi-party service, $1.50 for town multi-party residence, $1.75 for town one-party residence, $2.50 for town multi-party business, $3.00 for town one-party business, and $1.00 per month for extension phones.

Wallace Chance was hired in 1950 as superintendent of telephone repairs and switchboard maintenance.

"Wallace did an excellent job in training our new employees in telephone maintenance," Glenn says.

Chance resigned in 1957 to become a manager in Mayville, N. D.

In June, 1951, Don Jeffries was hired as supervisor of construction and heavy maintenance. He was assisted by Ray Wogen and Richard Stensrud in the Lake Mills area; Ray Hauswirth and Marvin Haukoos in the Forest City, Leland and Crystal Lake areas; and by Burnell Jordahl, with Jeffries, in the Rake, Buffalo Center and Thompson areas.

Jeffries, who had worked for the United Telephone Company in central Iowa, was an excellent employee and knew how to motivate his people, Glenn says.

In October, 1964, Jeffries resigned to go into the automobile business with a brother in Bend, Oregon.

Jeffries was succeeded by Marlin Senne, a young man from a farm in the area, who had been hired six years earlier by the RTA and who worked in construction, maintenance and dial switchboard service.

"Marlin was and is a very dedicated and ambitious employee," Glenn says.

Kenneth Lein was hired in August, 1951, as a member of the RTA's construction crew. "Ken was a young, bright, personable fellow," Manager Bergland says. "He progressed through the ranks during the next twenty-three years; first, in construction then in telephone maintenance repairs. Later, he became a switchboard technician and was moved up into the office to become a work order and inventory control clerk. Soon he was moved into engineering, then became a staff assistant. In 1974, he became manager upon my retirement," Manager Bergland added.

This 1951 photo shows Dale Jacobson pointing to the Co-op's coverage area following the consolidation of the first nine exchanges: Rake, Buffalo Center rural, Thompson, Scarville, Lake Mills rural, Joice, Forest City rural, Crystal Lake and Leland. Dale was an RTA work-order engineer.

Dale Jacobson was another employee who contributed to the success of the RTA in the early days.

Dale was a farmer west of Forest City and president of a small farm telephone company with 10 telephones, at the time of consolidation.

"Visiting with Dale at one of our organizational meetings, I learned that he had been a telephone technician in the army and had taken telephone technical studies at Georgia Tech. I told Dale he should be working for us rather than farming," Glenn relates.

"About a week later, Dale appeared in my office and asked me if I'd meant what I said. I assured him that I had, and we hired Dale as our work order clerk and engineer."

"Dale resigned in 1954 to be manager of a telephone co-op in South Dakota, and later left that job to establish his own engineering firm," (now in Estherville, Iowa), Glenn recalls. He later did considerable consulting work for the co-ops.

"The Cooperative in those early years was very fortunate in hiring very dedicated young people, who were interested in learning and were ambitious and personable," Glenn adds.

Indeed, back then, experienced telephone workers were not readily available. It was essential to the operation of the new cooperative that it provide training classes for its employees. The classes were taught by Glenn, and by Arthur Mitchell and Wallace Chance, supervisory employees of the RTA.

The classes were held at night during the winter, in the basement of the RTA office building. Subjects included basic electricity, basic electronics, dial switching and telephone repair and maintenance.

The RTA today also teaches some special classes.

"We also encouraged our employees to enroll in correspondence courses in practical electricity and telephony, all of which were paid for by the cooperative upon employee's satisfactory completion," Glenn says.

"We found that our best people were the local farm lads or rural town people. They were ambitious. We looked for ambition and a desire to work, as well as for honesty, an agreeable personality, and an ability to get along with people.

"By screening and training, we were able to put together some very fine telephone crews, of whom we were and are very proud, and they, in turn, can be very proud of what they have achieved.

"A great many continue to be employed by the Cooperative," Glenn says. "In fact, sixteen went on to become managers in other systems in the first twenty-five years."

Two Grandfathered Franchises

It was necessary for the cooperative, as a new legal entity, to secure franchises for the use of the streets and alleys in each of the towns served; and to secure easements from the owners of the farm properties on which poles and wires would be erected, or cables buried.

The law requiring franchises was passed October 1, 1897, and any town telephone system in existence prior to then had its franchise 'grandfathered,' i.e., it was granted a franchise without an election, and this franchise was granted in perpetuity—for the system and all its successors or assigns. All other franchises must be continually renewed in periodic elections, usually every twenty-five years.

The franchises for Lake Mills and Thompson in the Winnebago area were grandfathered.

As early as 1896, Thompson had a phone system, albeit a primitive one, of Dr. G. M. Lee (see story, page 3). Dr. Lee was still living in 1951 and in response to the RTA's request, gave testimony to a court reporter on the existence of the pre-1897 system, which testimony the RTA filed with the recorder's office. It is pictured on page 3.

The following history of the Lake Mills Telephone Company was researched by Glenn Bergland to use as proof that the Lake Mills franchise should be grandfathered.

Lake Mills Telephone Company

The November 4, 1896, issue of the *Lake Mills Graphic* carried a small item: "There is talk of organizing and perfecting a telephone system for Lake Mills. The scheme looks possible and no doubt will materialize."

The *Graphic* reported on March 24, 1897, that a mass meeting of the citizens was held the previous evening and a Mr. Thompson was appointed to solicit stock for a Lake Mills Telephone Company.

The company opened an office in the Massey building and construction began. The company's board of directors discussed the possibility of getting a long distance line from T. A. Way of Britt, who was a superintendent of the Western Electric Company, which had begun constructing long distance lines out of Mason City.

Two days prior to October 1, 1897, the town had its first telephone service, thereby becoming grandfathered under the new franchise law.

The *Lake Mills Graphic* reported that "the exchange opened for public service yesterday morning and many 'hello's' have gone through the wires. The central office is located in the S. P. Massey drug store and an efficient person is constantly in attendance at the switchboard. They claim to have the best "standard" telephones in the world and that you can hold a watch to any phone in town and hear it ticking in any other phone in town. You can stand in their office and talk distinctly over 500 miles of wire (this was questionable) and you can call the depot to make inquiries without wearing out 50 cents worth of shoe leather going down there."

The first farm 'company' started operations when a few farmers got together and chipped in enough money to run a line into the new switchboard.

In 1900, the switchboard was moved from the Massey building into the corner of the P. M. Joice building block. Articles appeared in the *Graphic* relating that the new companies were making necessary repairs and improvements to the service as a result of a great many complaints of poor service.

In the early 1900's, the Northern Telephone Company of Forest City purchased the Lake Mills Telephone Company.

In 1906, there was a failed effort by some to sell stock to form another telephone company. Also that year, the city council discussed the possibility of having all the telephone poles painted, to mask their dirty, bare, weather-beaten look. The main street was unsightly enough with 20 to 30 wires running between poles.

In 1908, a new telephone company was organized in Lake Mills, its owners threatening to start operating if Northern Telephone didn't improve its service. The new company did, however, buy the Lake Mills Telephone Company from Northern.

By 1914, the Company was in the control of M. O. Godtland, who, starting as a lineman, had accumulated the shares of the owners as they moved away or died.

For 10 years, during the decade of the 50's, the Winnebago RTA tried to purchase Lake Mills Telephone from Godtland, but he steadfastly refused to sell because he had a strong antipathy for, almost a hatred of, cooperatives. He was a contributor to the Tax Equality League.

Elmer Bergland and M.J. Rygh represented the Cooperative in its purchase attempts, but always were rebuffed by Gotland saying, "I'll never sell to a cooperative; I'll sell to anyone else, even to Glenn." Mr. Gotland was set in his ways.

Numerous companies and individuals were courting Godtland, hoping to purchase his telephone company, and a couple of times he almost sold to them.

Mr. Godtland, "Mog," as he was called, did like Glenn, and often called him to look after the "shop" when Mog went fishing.

The RTA board of directors, after several discussions, decided it would be better for Glenn to buy than some unfriendly outsider. Several times over the next three years, Mog proposed a sale, but nothing happened, until May, 1961.

Glenn had been in Washington, D.C., for a week and had decided to finally accept the position of Assistant Administrator of the Rural Electrification Administration, in charge of the Telephone Division.

On the Saturday of Glenn's return, Mog inquired, "Where have you been? I have been wanting to sell to you all week."

In one hour, the attorney, Richard Ramsey, had drawn a contract and it was signed.

"Mog was an unusual and different sort of fellow," Glenn says. "If I came to see him and was cussed out, I was in 'clover', but if he was subdued and melancholy, I'd leave."

Thus, Glenn, in 1961, became the owner of Lake Mills Telephone, and put on a third hat—as manager of Winnebago REA, as manager of Winnebago RTA, and, nights and weekends, as manager of the Lake Mills Telephone Company.

Jerry Chulpach was hired in 1961 to be the supervisor at Lake Mills; he resigned in 1965 to be the distributor of Ditch Witch Trencher Equipment in Iowa.

Marlo Durby, who had assumed a Manager's position in Dows, Iowa, the year before, was solicited to return to become exchange manager for Lake Mills.

"Marlo Durby was, and is, a very dedicated, intelligent telephone technician. He helped rebuild the Lake Mills exchange into a modern dial electronic system with all the features available in large city exchanges." Glenn says.

In 1982, Glenn sold the company and the TV cable company to the Winnebago RTA.

* * *

Franchises and Easements

The RTA was granted a franchise in each of the towns (Lake Mills and Thompson excepted). The procedure was relatively simple, but time consuming. Glenn, usually with the attorney, Dudley Weible, would meet with the town councils and request a franchise election. Forms and ballots would be printed (at RTA's expense), legal notice of the election would be published at the town newspaper, and the election held, all in a period of about six weeks. All of the elections were favorable to the RTA.

For both its new lines and the replacement of the old company lines, the RTA obtained rights-of-way and easements from owners of the farmland involved. With more than 2,000 farms served, it was a big job.

Easements fixed the right-of-way as the first five feet of farmland adjacent to the public road right-of-way. Poles were set or cables laid one foot inside the property line. If the road right-of-way is sixty-six feet, for example, the RTA line would be 34 feet from the road center. The surveys followed the county engineers' cornerstones, and many of these, in turn, dated back to the government's original surveys that fixed the boundaries of six-mile square townships, each township comprising 36 square mile sections, each section being 640 acres.

Note on north-south boundaries: These run parallel to meridians of longitude. The meridians converge at the north and south poles; consequently, going north there is a progressive decrease in the distance between two meridians, which are at maximum distance apart at the equator and zero distance apart at the north pole.

Consequently, there also is a progressive decrease of the *number* but not the size of the townships between any two meridians. The north-south borders of the northernmost tier of townships in Winnebago county do not align with their counterparts in the southernmost tier of townships in Freeborn county, Minnesota, because of the lateral offsets resulting from placement of the townships to accommodate the converging meridians.

* * *

Most of the easements were negotiated for the RTA by Charles Peckham. "We were very fortunate in having with us Charles Peckham, who had obtained right-of-ways for us for the REA from its beginning in 1938," Glenn relates.

Mr. Peckham had been an inspector for the Iowa Department of Agriculture and had retired to operate a grocery store in Thompson when the REA hired him on a part-time basis.

"After assisting the REA for a number of years, Charles moved to Carmel, California, but before long moved back to Thompson and built a home. We asked his assistance in our meetings with the various companies and in soliciting votes for us, and in signing up RTA right-of-ways.

"Mr. Peckham had a tremendous personality and was very effective. He was devoted to his work and to the cooperative, and his friendly manner made him welcome in every farm home," Glenn added.

The Cooperative did not pay for any of the easements; the landowners agreed with the Cooperative that the easements were for their benefit as Cooperative members.

Commentary on Dudley Weible

By Glenn Bergland

"The Cooperative was fortunate in its beginning to have as its attorney Dudley Weible, who then was known as one of the very best attorneys in northern Iowa.

"Dudley would battle for what he believed would be the right solution to our problems, to which he took a practical approach; his modus operandi saved us time and needless effort.

"The 67 legal entities that were consolidated into the telephone cooperative all required different legal procedures, whether they were small unincorporated associations of eight or ten people who actually owned the lines and from whom we had to get individual permission and signovers, or whether they were larger stock companies, cooperative or mutual membership companies.

"Dudley prepared a set of forms for the calling of the first board meeting of each association to authorize the election as to whether or not the membership wanted to consolidate, and then a different set of forms for the second election meeting, and yet another set of forms for the board of directors of an association to sign over its assets to the new consolidated company. These forms occupied sixteen feet of space along the wall in our boardroom.

Dudley Weible and Glenn Bergland at Dudley's retirement party, 1986

"At many of these meetings, Dudley was present to smooth out the legal obstacles and to help answer questions.

"Naturally, all of the regulations and legal procedures could not be followed in their entirety so as to be 100 percent legal without challenge, but Dudley took the position that they *would* stand up in court.

"At first, the REA legal division in Washington, D. C., wanted every 'i' dotted and 't' crossed, but we finally found an attorney in the REA administration who took a practical approach, the same as Dudley. Dudley was able to convince him that our papers were in order and that there would be no legal problem.

"Dudley did a tremendous job for us and a great deal of thanks goes out to him from the Winnebago RTA."

REC/RTA Joint Contract

The joint management contract between the Winnebago Rural Electric and the Winnebago Co-op Telephone Association utilized a great number of the electric office employees during the early years. They were actually the sole staffing available to the Telephone Association, so they had dual positions to fill. From the beginning, the two cooperatives were set up as separate legal entities, each with its own financial obligations and record-keeping, even though these functions were performed by the same people in the same office.

Palmer Hamlin, who was the REA office manager, kept the books of both the electric and telephone.

"Mr. Hamlin became very knowledgeable of the telephone operations and did a commendable job in assisting in the formation of the cooperative from 1949 to 1974, and when I resigned we separated the two organizations, Mr. Hamlin taking over the management of the REA, and Ken Lein the management of the RTA.

"Mr. Hamlin was very dedicated. He was an excellent accountant and received many accolades from the auditors of our books. Generally, we had three audits each year—by the General Accounting office of the government, by the REA itself, and by our own independent auditing firms. They all commended Mr. Hamlin for his excellent work," Glenn added.

Later, Mr. Hamlin was assisted by Eugene Sauer, who served also as the telephone accountant from 1954 until 1973, when Robert Houtkooper was hired to do the telephone accounting and Mr. Sauer was assigned to the electric accounting. Mr. Houtkooper suffered a fatal heart attack in 1981 and his position was filled by Terry Wegener.

During the early days, "Eugene Sauer did a tremendous task in keeping records for the two cooperatives. He was and still is very conscientious and sincere," Glenn says. "He was very dedicated and capable."

Dorothy Houg, REA stenographer and cashier/receptionist (who has since retired), had her workload increased tremendously when the RTA started. Dorothy answered questions not only on the electric, but the telephone, too, as well as helping the additional clients.

Board Members

"I was fortunate in having tremendous board members on the REA. They became the board of the RTA also (see page xi) and were most supportive and helpful in the establishment and growth of the RTA. They were dedicated crusaders who saw the need for better rural telephone service," Glenn says.

Elmer Bergland was the first president of the RTA.

"Not only did Elmer provide great moral support to the management, but he was very active in our many meetings, at home and away.

"Besides our trips to Cedar Rapids, related earlier, Elmer went with Dudley Weible and me to Washington. 'Went with'—well, not exactly. Dudley and I flew, but Elmer had never flown—these were the early days of airplane travel—and he was apprehensive of flying. He took the night train from Mason City and joined us in Washington two nights and a day later," Glenn recalls.

Glenn occasionally introduced Elmer as his "father," although the two were not related. Eyebrows would be raised because nepotism was frowned upon. On one occasion, the federal REA administration even sent a questioning letter stating that if Elmer and Glenn were related, one or the other must resign.

Elmer Bergland resigned the REA/RTA board presidency in 1953, when he moved to Lake Mills from his farm outside of Thompson.

Elmer was succeeded by R. E. Aukes. The other board members kidded Mr. Aukes as being "dumb . . . like a fox!" Feigning ignorance was Mr. Aukes' way of getting answers to questions.

"Mr. Aukes had an ability to cross-examine by raising questions on any issue that came before the board; even though he favored an issue, he would play the devil's advocate to bring out the facts," Glenn recalls.

Mr. Aukes was the "boards" president for 20 years, until he resigned in 1973. He became known as "Mr. REA." "He was a masterful chairman of the board," Glenn says.

Mr. Aukes farmed south of Buffalo Center. He kept a large flock of turkeys and as a hobby flew his own airplane and had a landing strip and hangar on his farm. To this day, he continues to fly his own plane, at age 92!

At the time this book was sent to the printer in June, 1990, Mr. Aukes was 92 years old and piloting his own airplane.

Gustaf Gustafson succeeded Mr. Aukes as president. Mr. Gustafson was a charter RTA board member. His father, Carl Gustafson, was a charter REA Electric board member.

"Mr. Gustafson was a stabilizing influence upon the board of directors. He sought to raise questions on the issues and generally came up with an intelligent and common sense solution. He was a valuable member of the board," Glenn states.

Mr. Gustafson was succeeded as president by Arlo Carson, who was appointed to the board in 1956 after the death of M. J. Rygh.

In 1976, Vernon Jacobson was elected president. Mr. Carson retired from the board in 1979.

"Mr. Jacobson has provided steadfast momentum to the modernization and expansion of the telephone cooperative," Glenn states.

"The Telephone and the Farmer"

In January, 1951, when RTA manager Glenn Bergland was in Washington to work out the details of the REA loan to the telephone cooperative, he learned that the REA was planning to make a promotional movie that winter to be used by the REA field staffs throughout America to encourage other rural electric systems to promote rural telephony like the Winnebago REA was doing.

The movie's producers planned to film it in northern Minnesota so as to include snow scenes and snow storms.

Glenn convinced the producers to film the movie in Winnebago county, for "after all, the movie is supposed to be a reflection of the Winnebago achievement, and besides, we can promise you that there will be a snow storm before the end of February," Glenn said.

Film director Irving Rusinow accepted Glenn's invitation and Glenn's assurance that the RTA would assist in the project; in fact, the characters would be played by local people.

Mr. Rusinow, Homer Boor, cameraman, and sound crews arrived in the county in February, 1951, and although there was little snow on the ground, within days there was so much snow that Mr. Rusinow was heard to say, "It's like a Christmas card, unbelievable."

"The Telephone and the Farmer" was the movie's name. Leading roles were played by Dr. Charles Potter of Forest City and C. E. Judd of Thompson, portrayed as county agents; Ray Murray of Buffalo Center as the "doctor;" Frank Murray of Buffalo Center as the country storekeeper; Mrs. Willard Sprecher of Scarville and Arthur Mitchell of Thompson as "Mr. and Mrs. Porter," "Mr. Porter" being the patient stricken with a heart attack; Ann Kruse of Thompson as the telephone operator; Palmer Hamlin of Thompson as a neighbor boy; Bruce Thompson of Forest City as the REA manager; and Mrs. John Benjamin and Jack Griffin of Buffalo Center as farmer neighbors of the "Porters."

Other participants included Don Jeffries, Charles Peckham, Charles Brooker, Johnny Efting and Art Ashpole, all of Thompson; Ed Johnson, Mr. and Mrs. Jens Jenson, Klaus Risius, Oscar Nelson, Bert Dearing and Mr. and Mrs. John Benjamin of Buffalo Center; Wilfred Peterson and Dewey Hanna of Lake Mills; and F. Buren, Leland.

Settings for the movie included the Dr. H. H. Murray office at Lakota, the Erickson store and the REA office at Thompson, the Clawson barber shop at Buffalo Center, the Lake Mills city hall, the Frank Murray and John Benjamin farm homes, and the Winnebago County Farm Bureau annual picnic at the Thompson park.

"The Telephone and the Farmer" had its premier showing on Saturday, December 19, 1951, at the Forest City civic auditorium. Claude R. Wickard, REA Administrator, was the guest speaker, introduced by Elmer Bergland, RTA president.

Mr. Wickard said that the 1950 census shows the percentage of farms having telephones is no longer than when the 1920 census was taken, and the deterioration in rural telephone service led to the REA Telephone Lending Act.

The Winnebago County Board of Supervisors proclaimed December 19 as "Rural Telephone Day." The official proclamation recognized the successful formation of RTA and the securing of the REA loan, and the premiere of the movie. Ernest Nelson, Supervisors chairman, signed the proclamation.

The movie showed the difficulties encountered by the farmer's wife in summoning medical aid by telephone after her husband suffered a heart attack, and portrayed, step by step, the organization of a telephone cooperative and the obtaining of a loan from the REA.

The movie was shown hundreds of times throughout the nation to inspire rural electrics to promote telephone cooperatives similar to the Winnebago RTA.

Alas, the commercial companies complained that the REA favored cooperatives, so in July, 1953, Ancher Nelson, REA Administrator, saw fit to revise the first 10 minutes of the film, which depicted the inadequate service pre-Co-op telephones; and then after more commercial complaints, the film was withdrawn from distribution.

The film copy in the Winnebago RTA office may be the only copy there is outside the REA vaults.

Helping the Shut-Ins

On numerous occasions in its early years of operation, the RTA made several school-to-home telephone installations for handicapped or disabled students who were home or recuperating.

Two such installations were for Jerry Thorland, a Thompson high school sophomore who, in 1952, was ill with rheumatic fever, and Gene Munson, a seventh grade student at Charlson rural school, who in 1951 also was ill with rheumatic fever. The Cooperative installed a radio line carrier system, via two-way radio waves that were superimposed on the telephone wires, between Jerry's and Gene's farm homes and their schools.

Radio telephone carrier is a means by which additional circuits can be added to one pair of telephone lines so as to transmit several telephone conversations at one time over the pair. There were speakers and receivers in the classrooms and speakers and receivers in the boys' bedrooms by which they could communicate directly with their classes and keep up with their studies.

Such installations were a combined project of the Iowa State Department of Education, the Thompson high school and the Winnebago Telephone Cooperative.

* * *

In the beginning of dial telephone service, many people had never seen dial phones and Cooperative personnel held demonstrations at area schools on the use of dial, and explained how dial functioned and how it would operate and actually do the switching for calls to neighbors and others in the country.

RTA Crews Save Dollars

From its beginning, the Winnebago RTA directly purchased all material used for lines repair and construction. Much of the construction of overhead lines was contracted out, but the RTA's own crews did all of the line rebuilding along the roads that the county re-graded and backsloped. These lines were moved onto RTA right-of-way on land adjacent to the road right-of-way.

The underground cable in the towns was laid by RTA crews, also, because of the great number of obstacles that would be found underground, which a contractor could not foresee and so could not include in a bid.

Beginning with the changeover to all underground rural cable, and ever since, the Cooperative has done most of its own construction and engineering with its own crews, at a savings to date of well over $600,000.

Aerial Construction

In the spring and summer of 1951, the RTA staff worked on detailed plans and engineering for the new dial system. Since very few independent telephone construction specifications existed, REA used Bell Company specs as a guide, with modifications.

Until polyethylene cables were developed in the mid sixties, when cables could to be buried underground at a reasonable cost, it had been financially necessary that the RTA build its first rural system with poles and overhead wires, with seven or eight parties on a pair of wires.

On August 3, 1951, after securing numerous bids and quotations, the RTA let the following contracts for material:

One-third of all pole requirements and all crossarms to Westinghouse Electric Supply Corp., Waterloo, Iowa, $30,008.95; all of the telephone lead cable and one-half of the telephone wire required, all of the terminals and protectors and one-half of the hardware required to Graybar Electric Supply Co., Des Moines/Chicago, $103,389.42; one-half of the hardware and one-half of the telephone wire to Crescent Electric Supply, Mason City, $32,666.66, and 11,000 feet of plastic cable to Terry-Durin Electric Supply Co., Cedar Rapids, $1,181.56.

Only a third of the required poles were ordered because poles were then in plentiful supply and the Cooperative could order on an as-needed basis.

On average, a telephone pole was 16 to 30 feet long, with three and one-half to five feet below ground. The RTA generally used long leaf southern pine poles. The poles were pressure and vacuum treated throughout their length with creosote, the REA standard being 8 pounds of creosote retention per cubic foot.

With overhead lines, it was not feasible to provide single-party service for rural patrons. Each single-party phone requires two wires running from phone to town switchboard. Fifty private-party phones required 100 wires. There could be just two wires on the pole for the farm farthest out, but for each single-party phone added, there would have to be two more wires, for an impossible total of 100 wires on the poles nearest to town. Such would be physically and financially out of the question. Rural single-party service became practical in the middle sixties with the advent of cheaper polyethylene plastic buried cable and radio carrier.

In the beginning, RTA provided multi-party phones, but with private-line ringing, so that when a person was dialed, only that person would hear the ring. This was accomplished by ringing generators with five different frequencies—20, 30, 40, 50 and 60 cycles. The ringer on a line, which was resonant to an assigned frequency, would be the only phone that would ring; and with wire-to-ground ringing on each of the wires in the two-wire circuit, ten different rings could serve a ten party line.

Dishpan Alert

It is surprising how innovative people are. Before RTA, when one phone on a party line rang, all the phones on that line rang. Then others on the line would pick up their phones and listen in, notwithstanding it was not their special ring. Did RTA's private ringing stop the clandestine listening? Not quite! Many households discovered that if they put their telephones on a dishpan, there would be an audible vibration when another party was called. When the vibrating stopped, the curious would pick up their phones for a bit of listening in. The RTA maintenance men found many a telephone on the dishpan.

On October 3, 1951, the Cooperative accepted a bid of $97,000 from the Cushing Construction Company of Woodbine, Iowa, to build approximately 200 miles of rural overhead telephone lines for the Thompson, Rake and Joice exchanges.

Five contractors had submitted bids. The highest was $185,000, the next high, $129,000.

Work scheduled for the RTA's own crews included the following:

Build approximately 58.7 miles of new telephone lines along the new road grades scattered throughout the areas; rebuild 53.8 miles of line and 500 drops in various areas where the line can be repaired to A-1 condition; rewire 2,179 houses for telephones and install the new dial phones; install approximately 27 miles of underground cable in the towns and rural areas where heavy leads extend out of the towns; rebuild the outside lines in the towns of Rake, Scarville, Leland, Joice, Thompson and Crystal Lake; build new buildings in Rake, Joice, Scarville, Thompson and Leland to house the new dial switchboards; do the engineering work in the preparation of plans, specifications and priority applications for the entire program; and span and stake 750 miles of line, i.e., determine distance between poles and mark with stakes.

Three men could span and stake and prepare staking sheets for about three to five miles of line per day.

A contract was awarded in November, 1951, for $96,000 to Graybar Electric Company for the installation of North Electric automatic dial switchboards, for the exchanges of Rake, Thompson, Scarville and Leland, with delivery and installation scheduled for the fall and winter of 1951-1952.

"This means that most of these telephones can be cut over to dial in a month or two after switchboard installation," Manager Bergland said.

Briefs - 1951

R. E. Hauswirth, 42, RTA employee, died June 26 after suffering a heart attack while repairing tornado damage at 2 A.M. to telephone lines at Crystal Lake.

—

The November 29 issue of the Forest City Summit pictures RTA employees Kenneth Lein, Don Jeffries and Paul Batton sitting by the bedside of Gene Munson, 7th grade student communicating by radio phone with his Charlson rural school.

—

On June 1, the RTA began sending telephone bills to the 2,300 former members of the old telephone systems that now are part of the new RTA.

—

In his speech October 19 at the Forest City civic auditorium at the premier showing of the movie, "The Telephone and the Farmer," REA administrator Claude R. Wickard said that the 1950 census shows the percentage of farms having telephones is no larger than when the 1920 census was taken, and the deterioration in rural telephone service led to the REA telephone lending act.

—

In November, 1951, the Winnebago REA/RTA inaugurated its own training school for REA and RTA linemen and RTA installers. It was one of the first REA schools of its kind in the state. Included were lessons on the nature of electricity, magnets, magnetism, electromagnetism, currents, Ohms law, electric circuits, batteries, generators, transformers, motors, meters, power facts, inductance, capacity and static electricity. Since the above subjects are common to both electric and telephone crews, they attended as a group for the 17-week course, after which they divided for training on their own. The instructors were Arthur Mitchell, REA employee, and Manager Bergland.

1952

In January, 1952, the RTA awarded a contract to Felix Halverson, a local contractor, to build switchboard buildings in Rake, Thompson, Scarville and Joice. (Of wood frame construction, these first switchboard buildings were replaced in later years with concrete block structures to house the private line switchboards installed in the 60's.)

In March, 1952, the Cooperative awarded a $180,633.07 contract to the Harrison Construction Company of Hampton, Iowa, for the construction of telephone lines in the rural areas of the Lake Mills, Scarville, Leland and Forest City exchanges.

Unsuccessful bids were $248,555, $208,291, $198,007 and $194,453.

The bid figures above reflect the astute business management that characterizes the Cooperative. Time and time again in the history of the Winnebago RTA (and the Winnebago REA) the prudent and dollar-watching management of Glenn Bergland is evident, as are these traits evident in his successor, Kenneth Lein and his associates, many of whom were trained by Manager Bergland.

By February, 1952, material and equipment prices had increased by an average of 24 percent over those used for the original loan application. To provide for the cost increase and also for an additional 337 subscribers, the Cooperative applied to the REA for a supplemental loan of $372,000, which was granted in May, 1952.

By July, 1952, the Cooperative's own construction crews had installed more than 600 new dial phones, cleared trees on 72 miles of line, set poles on 66 miles of line, strung wire on 65 miles of line, laid six miles of underground cable, and temporarily repaired 53 miles of line on which construction would be delayed for up to three years pending the area's conversion to dial.

December 30, 1952! A Historic Date in the Annals of the RTA

The Rake telephone exchange became the first RTA cooperative system in Iowa; the *third only* in the nation!—to be "cut over"— converted to the new, modern dial telephone service with the use of REA loans.

It demonstrated the truism, "where there's a will there's a way." Manager Bergland and the RTA board of directors *willed* that there be a better telephone service for the Winnebago citizens. Under the superb organizational ability and leadership of Glenn Bergland, there was a *way*: Get the support and cooperation of the community, secure an REA loan by marshalling the facts—having one's ducks in order— and starting from scratch quickly and efficiently build a brand new, modern rural telephone service.

Third in nation! Cutover to dial of the Cooperative's Rake exchange on Tuesday, December 30, 1952 was the third only such dial cutover in the nation by a rural telephone company. Rake's mayor, Frank Delyea made the first call, to REA Administrator Claude Wickard, in Washington. Looking on are, left to right, RTA maintenance foreman Wallace Chance, maintenance technician John Osnes, and Manager Bergland.

Rake's mayor, Frank Delyea, made the first call from the dial exchange to the REA administrator in Washington, Claude R. Wickard. Mr. Wickard stressed that the government "acted as a banker, that the organization, interest, control and management rested with the members of the Winnebago Cooperative Telephone Association."

Manager Bergland then talked to Mr. Wickard, who told Glenn, "You folks still have a record here in REA which has not yet been broken. There have been many consolidations of telephone companies to obtain loans from us, but none has equalled the 44 consolidations which were effected to form the Winnebago RTA."

Today's Winnebago RTA is a consolidation of 69 companies.

In one hour following the cutover, 950 calls were made or attempted by the 243 Rake subscribers.

Briefs - 1952

The Cooperative is in constant touch with the National War Production Authority, similar to the War Production Board of World War II, for authorization of priorities for the purchase of steel, copper and all materials.

—

People will see about as much difference between the old telephone service and the new dial service as they saw between kerosene lamps and electric lights, the RTA said.

—

Dudley Weible, newly elected State Commander of the Iowa American Legion and who is the RTA attorney, spoke at the Cooperative's annual meeting March 19.

—

Herb Copeland and Don Gray, RTA telephone installers, advise that installation will be speeded up considerably if patrons will give advance notice of choice of wall or desk phone. Price for each is the same.

—

All of the Crystal Lake subscribers had dial phones by the end of July, but final conversion of the exchange was awaiting the arrival of the new automatic switchboard.

—

New Switchboard buildings for Thompson, Rake and Joice were completed in July.

—

The new dial numbers for Rake have been assigned, and the directory is at the printer. It will be ready by the time of the Rake cutover.

—

"We have had more trouble than should be expected in the Scarville, Lake Mills, Leland and Forest City areas," the RTA announced. "This is because of the failure of the Harrison Construction Company to provide sufficient repairs for continuity of service on the old lines, so as to not cause trouble during or after construction."

—

Charles Peckham is ninety-nine percent finished getting right-of-ways for the entire RTA system.

Filming "The Telephone and the Farmer." Bruce Thompson, Forest City, left, played the part of REA manager, and Ray Murray, Buffalo Center, right, that of county agent.

This setting for "The Telephone and the Farmer" was on the second floor of the Lake Mills City Hall.

1953

In early January, 1953, the Joice exchange was cut over to dial. Joice Mayor Burton Aasland made the first call, to his sister, Mrs. Richard Snyder, in Los Angeles.

The new lines in the Joice rural area were built by the Cushing Company; the Cooperative's own crews did all the construction in the town of Joice, which included the laying of 6,800 feet of buried telephone cable.

On March 12, the Leland exchange was cut over to dial, Mayor W. P. (Hap) Buren making the first call, to his niece, Mrs. Patrick Stokes, in Seattle, Washington.

On March 31, the Scarville exchange was cut over to dial, Mayor Melbourne Bye making the first call, to his father, Selmer Bye, in Hillsboro, Oregon.

Following the Cooperative's policy of enabling neighbors to talk to neighbors where feasible, subscribers can make calls without charge to subscribers in exchange area contiguous to their own; Scarville subscribers now could make free long distance calls to all the Lake Mills and Thompson subscribers, town and rural, etc.

The policy, called Extended Area Service, still prevails today. Lake Mills subscribers can call free to Leland, Joice and Scarville; Joice subscribers to Fertile, Forest City, Hanlontown, Kensett and Lake Mills; Forest City rural to Crystal Lake, Joice, Fertile, Leland, Miller; Crystal Lake to Forest City, Thompson, Britt, Woden; Buffalo Center to Rake, Thompson, Woden; Kensett to Grafton, Joice, Manly, Northwood; Rake to Buffalo Center and Thompson; Thompson to Buffalo Center, Crystal Lake, Leland, Scarville, Rake; Twin Lakes to Albert Lea, Conger, Emmons; Woden to Crystal Lake, Britt, Buffalo Center, Titonka, Wesley; Fertile to Forest City, Hanlontown, Joice; Emmons to Albert Lea, Conger, Twin Lakes; Grafton to Kensett, Manly, Plymouth; Hanlontown to Fertile, Joice, Manly; Leland to Forest City, Lake Mills, Thompson; and Scarville to Lake Mills and Thompson.

On May 13, 1953, the Thompson exchange was cut over to dial; Crystal Lake was cut over in late summer; Forest City rural in July; Lake Mills, in September, and Buffalo Center in December.

Crystal Lake was cut over in late summer, Forest City rural in July, Lake Mills in September, and Buffalo Center in December.

Thus a milestone was reached in the history of the Winnebago RTA. The nine exchanges named in the original REA loan application had been converted to dial. They were: Buffalo Center rural, Crystal Lake, Forest City rural, Joice, Lake Mills rural, Leland, Rake, Scarville and Thompson.

In each home just prior to cutover there were two phones. One was the new dial phone, which wasn't operable until cutover. The other phone was the old magneto phone, which required a brisk cranking to either ring the operator if a call was to be made to someone not on the party line, or to ring directly to a neighbor on the party line. Many of the old replaced phones were the property of the subscribers, who pitched most of them, oblivious, alas, to their future value as antiques.

The old phones that were owned by the Cooperative were also mostly pitched; some were sold to a dealer in California.

The new dial phones were owned by the Cooperative.

The old telephone poles and wires that were replaced by new lines were given to the property owner, with the stipulation that he dismantle them within a certain period of time, usually 30 days, after which the Cooperative sold or junked them.

Briefs - 1953

The heavy snow on New Year's Eve broke more than 400 poles of the old telephone lines and tangled many of the lines. The Cooperative's telephone construction crews were called out New Year's Day to fight the trouble with the help of the regular telephone maintenance crews.

—

President Elmer Bergland gave the following message on the occasion of his resignation from the REA/RTA board: "It is with reluctance and a feeling of regret that I sever connections with your fine organization.

"In doing so want to express my thanks and appreciation to the management and all employees for the courtesy and respect accorded me; to the board of directors for their loyal cooperation; to Weible and Slife for their sound legal advice on the important issues which came up from time to time. Last but not least, to the members for the privilege of working with you and for you.

"My best wishes to my successor, to the REA and the RTA."

—

The free trunk line between Thompson and Rake was put in service. Subscribers in Thompson first dial 8 and then continue with the number they are dialing in Rake. Rake subscribers first dial 8 and then continue with the number they are dialing in Thompson.

—

On August 10, the 273 subscribers of the Emmons Telephone company cast 229 votes in favor of joining the Winnebago RTA. The addition of Emmons' 550 phones brings the number of RTA phones to more than 3,000.

—

Following is an excerpt from *Wallace's Farmer* magazine, April, 1953:

"Manager Bergland points out that rural telephone service is not attractive to private investors. Increased use does not bring increased revenue, as in REA. The big towns are the cream with their high usage toll lines and high rentals. To hold down the cost, a rural system must be large.

"How big. He thinks their system, which appears big with its 2,500 phones, still is not large enough. It could handle 1,000 more with only one or two more on the force, he says. Any system must have a good manager, well kept records, and a good foreman. The intricate dial equipment requires little servicing. But when it does need servicing, it takes an expert, he said."

(The RTA today has 6,250 access phone lines.)

—

Local phone conversations are automatically limited to about seven minutes. A warning signal comes on the line about one minute before the line is cut off. "I think that's all right," said Mrs. Irvin Slattum. "One shouldn't monopolize the line too much."

1954

In May, 1954, Manager Bergland was appointed by REA Administrator Ancher Nelson to serve on the telephone advisory committee of the national REA.

In June, the National Rural Electric Cooperative telephone committee, of which Manager Bergland was chairman, met in Washington, D. C., with Mr. Nelson and his staff. Problems addressed included the preparation of uniform extended free service and a toll contract between REA borrowers, the Bell Telephone Company and independent companies, and the reduction of equity contributions by telephone patrons to secure loans. (The latter problem did not arise with Winnebago County because the Cooperative was able to secure loans without member contributions.) Other topics were construction specifications and construction costs, and the formation of a national telephone association of the REA-financed telephone cooperatives.

On October 6, 1954, the Winnebago RTA received approval for a $261,000 REA loan for converting the Emmons, Conger and Twin Lakes telephone systems to dial. Six hundred phones will be installed, 179.4 miles of telephone lines rebuilt and seven miles of telephone cable laid.

In November, 1954, the Cooperative applied for a telephone rate hearing before the Minnesota rate commission, which sets rates based upon fairness to both the consumer and the investor. The application was for the Emmons and Conger exchanges. (A favorable hearing was held on February 9, 1955.}

Briefs - 1954

The August *Electrik* reports that "there has been an unusual amount of lightning the past few weeks, resulting in hundreds of blown telephone fuses. Some of the subscribers have hesitated to report their telephones out of order right away because they felt that the telephone maintenance men were 'snowed under' with telephone reports. We would prefer to have you report your telephone trouble as soon as possible in order that we can better analyze the extent of the trouble and clear all the trouble in a particular area while the men are in that area."

—

Nearly 3,000 people visited the new REA/RTA building in Thompson August 14 and 15. The overflow crowd wound at times for nearly a block as people waited for guided tours of the air-conditioned office building and shops. There were free orchids for women and free coffee and donuts in the cooperatives' new dining room. The new headquarters was built of necessity as the result of the growth of the rural electric and telephone cooperatives. The first "office" of the Winnebago REA Cooperative was a single desk in the corner of the Farm Bureau office in Thompson.

—

The November 9 issue of the *Mason City Globe-Gazette* pictures Manager Bergland holding a radio telephone. The line beneath the picture reads: "The Winnebago REA/RTA radio communications system links by radio telephone a dozen trucks and cars and saves hours of needless driving. The North Iowa cooperative has received national recognition for its progressive ways."

—

In December, the 18-member REA Telephone Advisory committee to the REA's rural telephone loan program met in Washington to seek solutions to the half dozen or more pressing problems in the loan program. Manager Bergland is a member of the group, which includes representatives from independent and cooperative telephone associations, government agencies, AT&T, and telephone consultants, nationwide. The agenda included the words, "in the interest of encouraging the continued improvement and expansion of telephone service in rural areas, and to assist this program generally . . ." Topics included financing, bidding, equity, cooperation and extended area service.

—

The August *Electrik* pictures a "telephone display" operated by Ken Lein, construction lineman, John Osnes, maintenance man, and Wallace Chance, maintenance foreman.

—

From the "Manager's Corner" in the October, *Electrik*: "One hot afternoon I enjoyed the comforts of an electric room air conditioner at Mr. and Mrs. Klaas Meyer's home. In a couple of years the number of air conditioning units in farm homes may be as prevalent as electric clothes dryers."

Also, quoting Manager Bergland: "Read an article in *Successful Farming* magazine entitled 'The Case for Farm Price Supports.' Irrespective of the arguments concerning rigid vs. flexible support prices there was one statement that made sense: 'Many billions in subsidies have gone to business and industry. A House Appropriations Subcommittee in January, 1954, published figures indicating subsidies amounting to over forty-five and one-half billion for business since World War II—a large part of this for business conversion payments. In contrast, farm price support and surplus-removal operations in the last twenty years have cost only three and one-half billion or one percent of the value of crops and livestock marketed.' Note that the figure for business is only for the period since World War II, while the farm-program figure is for the past twenty years."

—

The Rake school (Sup't C. W. Porter) set up a vocational instruction shop in the former creamery building, and the RTA installed a telephone intercom system between the shop and school building.

This dated photo shows the RTA's Foster Hartman on a telephone pole checking a terminal point connection of pole wire with underground cable.

1955

In June, 1955, a contract was awarded to Stromberg Carlson Company for the installation of dial switchboards at Conger and Emmons, completion scheduled for January, 1956. The Twin Lakes exchange was modified and expanded.

Glenn Bergland and Ken Lein. (1955)

Briefs - 1955

RTA assistant manager Dale Jacobson has accepted the position of manager of the Sully-Buttes Rural Telephone Cooperative in Miller, South Dakota. Mr. Jacobson joined the REA/RTA in 1950, as an engineer, and was appointed assistant manager in 1952 in charge of engineering and administrative work for the RTA.

—

Glenn Bergland was re-elected chairman of the National Rural Electric Cooperative Association's Telephone Committee at the NRECA's annual meeting February in Atlantic City. Glenn was also elected Vice-president of the National Cooperative Telephone Association, which met at the same time in Atlantic City.

—

The Cooperative advises procedure on fire reporting: "If you live in telephone exchange 'A' and belong to town 'B' fire department, the number one question seems to be, 'What to do if the free service lines are busy between 'A' and 'B' towns?' You may proceed as follows: (1) If someone is on your own line, you may 'butt in' and ask them to hang up for a fire emergency. (2) Then all hang up, so you can pick up the phone again and get dial tone. (3) If free line is busy, dial '0' operator, tell her you wish to place an emergency fire call to town 'B' fire department and give her their fire number."

—

The grand attendance prize at the REA/RTA annual meeting was a Frigidaire range, won by Mr. and Mrs. Burton Kenyon of Thompson, and the mail ballot award, a Frigidaire electric dryer went to Kenneth Hage, Scarville.

—

Manager Bergland writes in the *Electrik*: "How would you like an electric oven that will bake a potato in 2 1/2 minutes, cook an eighteen pound roast in forty minutes? You can place a hamburger on a china plate, the hamburger will be done in seconds, but the plate will still be cold. Unbelievable? No. It will some day be common in our kitchens—the microwave!"

—

RTA telephone crews worked all day Sunday, December 5, to restore telephone services disrupted by an ice storm. Several poles were downed by the rain and sleet. The wires had frozen together so the men in many cases had to "walk the lines" to knock the ice off to untangle the lines.

1956

The Conger exchange was cut over on January 16, 1956, Mayor Donald Babbitt making the first call, to a friend in Lenexa, Kansas.

Emmons was cut over January 25, 1956, Mayor John Benson making the first call, to his mother, Mrs. John G. Benson in Spencer, Iowa.

The Twin Lakes, Minn., dial phones were switched through the Emmons exchange.

For rate uniformity throughout its system, the Winnebago RTA adjusted all its Iowa telephone rates on July 1, 1956, to be the same as established by the Minnesota Railroad and Warehouse Commission for the Emmons and Conger exchanges. The adjustments were slight and increased rural service eleven cents; four-party residence, thirteen cents; two-party residence, nine cents; one-party residence, five cents; two-party business, twenty-nine cents; one-party business, twenty-seven cents; and multi-party business, five cents; plus tax.

Arlo Carson, Thompson, was elected to the REA/RTA board of directors in December, 1956, to fill the vacancy created by the death of M. J. Rygh. Mr. Rygh had served on the board since the inception of the REA in 1938.

Mr. and Mrs. Carson and twin sons, Ronnie and Robbie, age five, were pictured in the *Electrik* for January, 1957.

Briefs - 1956

Ancher Nelson has resigned as Administrator of the National REA to run for governor in his home state of Minnesota. Mr. Nelson succeeded Claude Wickard as REA Administrator.

—

She Named "Dial Tone"

Mrs. Walter (Greta) Aukes of Woden was the winner in the contest to select a name for the new Winnebago RTA newsletter. *Dial Tone* is the name. The judges commented that "there were many interesting entries. Quite a few, however, were suggestive of telephone gossip or chatter. It was the feeling that a majority of our subscribers would rather have another name.

—

Marjorie Judd, daughter of Mr. and Mrs. C. E. Judd, rural Thompson, gave specific honor to the Winnebago REA and RTA in a term paper she wrote for an economics course at Luther College, Decorah. Marjorie praised the role of cooperatives in the American free enterprise system. She wrote that "they are a stabilizing influence in this competitive nation and help level off prices and services to their members as well as the community as a whole. Cooperatives are associations of people. Their motivation is a non-profit service. They do not make a profit but a savings. . . . Farmers are the very backbone of our free enterprise system. The cooperative is a large part of the farmers' business."

—

Twice a week and on Saturday morning, the Cooperative employees attend their own school on basic electricity, taught by Manager Bergland and Art Mitchell.

—

In the August *Dial Tone*, the Cooperative explains how it finds the exact location of a lightning burn or moisture leak in a three-mile stretch of cable laid 30 inches in the ground:

Most of the time it is quite easy, but does take some time. First, we disconnect all the phones of the pair of wires in trouble, and then try to locate a pair that is not in trouble. A "Wheatstone Bridge" is connected to both pairs and resistance readings are obtained. Each size of cable has a certain number of ohms resistance per foot. The readings can be converted to feet, and we dig down to the cable. Usually an AC tone is put on the pair and then a pickup coil is passed over the cable to determine the differences in volume. This more or less confirms the previous test before the cable is opened. (The new electronic locaters have made the wheatstone bridge obsolete.)

—

The RTA's eastern area service man, Herbert Copeland, who was stationed in Lake Mills, found it necessary to move to a warm climate because of Mrs. Copeland's health. The family moved to Ft. Meyer, Florida, on September 22.

—

The RTA reports that the Emmons-Conger people phoning between the two places two out of three times will possibly be talking over radio carrier waves hitchhiking on the pair of wires strung between the two towns. "It is another new development we have adopted."

1957

At the Cooperative's annual meeting in March, 1957, the following Board of Directors' officers were elected: R. E. Aukes, President; M. J. Madson, Vice president; H. D. Hartman, Secretary-Treasurer.

As of September, 1957, the Cooperative comprised eleven exchanges with 3,100 telephones.

Private-party service for farms was somewhat expensive in the years before buried cable. Remember, each private line required its own pair of wires running to the town switchboard. In 1957, the cost of private-line service was $4.50 per month for residence one-party service, PLUS thirty-five cents per month per one-eighth mile from city limits. For the rural patrons served by the Buffalo Center, Forest City and Lake Mills exchanges, the per-month cost for private lines was $8.20 for residence, $9.20 business, plus thirty-five cents per one-eighth mile. The higher cost was because of the town companies' switching charges.

A peculiarity of long distance rates was (and still is) that out-of-state calls, which were classified as 'long haul', may cost less per mile than in-state long distance, depending on rates approved by the FCC or the respective state commissioners.

The following is excerpted from Manager Bergland's report for the year 1957:

> Ten additional lines were added in Emmons, 30 additional lines in Joice and six extra link circuits. Extra free service dial trunking equipment was added in Rake and Thompson, and extra lines added from Buffalo Center to Rake and Buffalo Center to Thompson.
>
> We have been retransposing and resagging two-wire lines to reduce the number of line troubles. Crossarm leads are being resagged.
>
> Toll use has increased terrifically. You enjoy a great amount of 'free service' lines. Do you realize the number of times you use them during a month? Have you attempted to figure how much more your telephone bill would have been if you paid toll on all of these calls?
>
> We are constantly training our employees on dial equipment, trouble shooting, public relations and how to do their job better and more efficiently. Almost all our employees have grown and progressed with the organization,

and we are proud of the achievements they have made. Our financial record is good. 1957 was our best year.

Your association has paid $86,915.42 on interest and principal on its loan. To date we have invested $114,984.58 in U. S. bonds for future replacements and storm fund.

In a few years, we will have to readapt all our dial switchboards for two-letter, five-number dialing.

Sincerely yours, Glenn Bergland, Manager.

Briefs -1957

Eight boys of Winnebago County, accompanied by Ed Scherich, Thompson voc-ag instructor and Art Mitchell of the Cooperative, attended the Sixth Annual Cooperative Institute at Memorial Union, ISU, August 5-6. The boys were: Erlyn Olson, Russell Tordoff, Glenn Anderson, Jon Simonson, Rodney Thorland, Roger Olson, Edgar Scherich, Carroll Gustafson and Lyle Johnson.

—

Wallace Chance, equipment and cable man for the RTA the past five years, has resigned this position to become district manager of the Pioneer Telephone Company's system in and around Hillsboro and Mayville, North Dakota. The company includes 11 dial exchanges and 5,000 telephones. The Chance family will move to Dakota June 1.

—

The *Dial Tone* reports that "on May 15 someone shot a .22 rifle bullet into our aerial cable south of Twin Lakes. It shorted out 26 telephone pairs and disrupted service for all of the Twin Lakes area and some areas east of Emmons until our men could locate the bullet in the four-and-one-half miles of cable and make a repair splice.

"Service was also disrupted in the town of Forest City when someone dug a trench and cut through an underground telephone cable with 400 wires in it. Many of these wires provided connections to our subscribers northeast of Forest City."

1958

In January, 1958, Norman Larson was appointed to the Winnebago RTA Board of Directors to fill the vacancy created by the death of H. D. Hartman. Mr. Hartman had served the board since 1938.

Mr. Larson farmed one mile west of Joice.

The following account of the Cooperative's installation in 1958 of two-way radio-telephone communication between home and school is of interest because it is an example of the technical leadership, state-wide and nation-wide, of the Winnebago RTA:

> The Winnebago RTA recently installed a unique radio carrier school-to-home communication system between the Buffalo Center school and the Harland Johnson farm six miles northwest of Thompson. This unit was installed to provide Ruth Johnson, a fifth-grader, with communication with her school room. Ruth is convalescing from an illness, and without the use of school-to-home communication, she might have missed a year in school.
>
> Previously, these installations have been provided by stringing two extra wires. The Winnebago RTA proposed a less expensive way—the use of radio carrier. Don Potter of Algona and Earl Fischer of Des Moines, from the Iowa Department of Education, also became interested in this new approach and prevailed on their department to work with the Winnebago RTA on a trial installation.
>
> Here is how this new radio carrier works. In the Buffalo Center fifth-grade room a rewired intercommunication unit was installed. The voice is carried over two wires to a radio carrier unit installed in the dial switchboard building. The voice waves are changed to radio frequency waves, which are transmitted over one of the Buffalo-Thompson free service lines. At Thompson, a radio frequency bypass was installed to transfer these waves to Harland Johnson's rural telephone line. At Johnson's there is a receiver which changes the radio waves back to a voice frequency, which is fed into another intercommunication unit and the voice is further amplified so Ruth can hear well. The Buffalo Center radio carrier unit has a 12 KC transmitter and 24KC receiver. At Harland Johnson's a 24KC transmitter and 12KC receiver were installed. Ruth can hear her teacher and the other pupils recite. Her teacher can ask her questions and Ruth can recite back to the school room.

> The remarkable part is that during this school-to-home communication by radio carrier over these telephone lines, there can be two other telephone conversations going on at the same time, one on the Buffalo-Thompson free service line and one on Harland Johnson's rural telephone line out of Thompson. None of the three conversations interferes with the others.
>
> The State Department of Education said this was the first time radio carrier had been tried on a school-to-home installation.
>
> The Bundleman radio corporation, which furnished the technical equipment on a loan basis, said in a letter to Manager Bergland that "to its knowledge, this is the first time that classroom telephone service has been attempted with our PLC equipment. You have every right to be very proud of helping the youngster."

Radio carrier telephone service for regular patrons became fairly prevalent in the Cooperative's telephone system in the late '50's because it was cheaper in many cases to install radio carrier than to build new telephone lines to furnish private lines out in the country.

Later, the RTA began plowing in (burying) plastic cables and generally had spare pairs for private lines. If not, radio carrier now became prevalent at reduced prices, whereby extra voice paths could be obtained from each pair of cables.

The Cooperative's financial statement for 1958 showed revenues of $189,817.10, which included interest income of $4,333.10; and operating expenses of $76,200.52, for gross income of $93,616.58. Deducting from this reserves for depreciation of $48,700.19 and interest paid and accrued on the REA loan of $19,971.89 left a positive balance of $24,944.50, which was applied to the REA loan.

Briefs - 1958

The Cooperative provides one-party radio carrier telephone service to Dr. L. M. Forland, veterinarian, who lives five miles east of Joice, and who previously had been on an eight-party line. The Cooperative also provides one-party radio carrier telephone service to the Mansfield store, three miles south and four west of Conger, Minnesota.

—

Miss Marjorie Seymour, Conger, Minnesota, telephone Colfax 5-2415, and Miss Karen Powers, Crystal Lake, Iowa, telephone 4724, won a teen-age telephone essay contest and were awarded new Ericofon telephone extensions. Miss Seymour wrote reasons why she wanted a telephone extension in her house; Miss Powers won for naming the advertising page in new directories, "R-Teen-Agers."

—

Party line manners printed in the September *Dial Tone*: We share our party line by keeping our conversations short. There's a trick to calling a long list of people without monopolizing the line—we don't try to call the whole list at one sitting. By making a few calls and then waiting a little, we give someone else a chance to call in or out on the line. We avoid starting in on a list during the early part of the evening when our line is usually busiest. Leaving the receiver off the hook by accident keeps the line out of use so no one else can call in or out on a party line. We respect the privacy of our neighbor's calls. If our neighbors who share the line lift their receiver and ask to make an emergency call, we hang up to clear the line for them. We know they will do the same for us.

Here is a joint use telephone/electric pole. The two wires slanting top left to lower right are 7200-volt electric lines. Parallel to them are two telephone lines. The balance of the wires, lower left to top right, are telephone lines. Used in road crossings, the combination assured safe clearance between the telephone and electric wires.

1959

Glenn Bergland was elected president of the National Cooperative Telephone Association at its annual meeting Feb. 5-7, 1959, in Washington, D. C. Glenn was a charter member and organizer of the NTCA, whose membership comprises the telephone cooperatives in the nation.

In his new role as president of the NTCA, Glenn appeared before the U. S. House and Senate Agricultural committees to testify in behalf of appropriations for the REA electric and telephone programs; specifically, that the REA set aside $35 million for the sole use of telephone cooperative borrowers. Heretofore, up to seventy-five percent of the REA loan funds were apportioned to commercial telephone borrowers.

"We have no way of knowing how much other REA telephone borrowers will need," Manager Bergland said. "Therefore, in order to be realistic and in order that we don't ask too much or too little for the entire program, we would like the amount ($35 million) to be set aside for cooperative type borrowers."

Program speakers at the NTCA meeting included members of the U. S. House and Senate; Clyde T. Ellis, manager of the National Rural Electric Cooperative Association, and David A. Hamil, REA Administrator. Both congressional speakers and Ellis defended the two percent interest rate then paid on REA telephone loans, Ellis saying that any increase in the interest rate "would completely kill off the telephone co-ops."

Hamil, however, said that "the day will come when the public will ask you to pay the going rate for money, an interest rate more in line with the cost of money to the government."

A meeting that same week of the National Association of Rural Electric Co-ops was also attended by Manager Bergland. Senators John F. Kennedy and Lyndon B. Johnson spoke to the group and both opposed any increase in the two percent interest rate and vigorously opposed any form of private financing, the idea for which had been advanced by President Dwight Eisenhower, who also spoke to the group.

RTA President Aukes and the board member Elmer Seim accompanied Manager Bergland to the meetings.

At the Winnebago RTA's annual meeting in March, 1959, the board of directors re-elected R. E. Aukes, President; M. J. Madson, Vice President; and Gustaf Gustafson, Secretary-Treasurer.

The board's report to the membership included the statement that "we are definitely opposed to a utility commission in Iowa. Being that cooperative users are owners, then as owners we regulate our own company; why pay extra money to an outside body to regulate yourself? Iowa rates are comparable to surrounding states." The reader is referred to page 82 for Manager Bergland's work in support of this statement.

In May, 1959, Glenn Bergland, president of the National Telephone Cooperative Association, sent out a "National Board Letter." It is reprinted here verbatim because it is illustrative of the philosophy and function of the NTCA:

National Board Letter

By Glenn Bergland, President NTCA

"As the newly elected president of the National Telephone Cooperative Association, I welcome this opportunity to reach all of our members and the non-members through this publication.

"NTCA was organized five years ago by the telephone cooperatives in the nation to help protect our rights and serve our needs. I feel we have done a good job up to this point. However, we still need more members. This year our goal is a 50 percent increase in membership and I personally wish to solicit your assistance in helping us reach this objective. If your neighboring telephone co-op does not belong to your association, ask him to join. NTCA is the only national telephone organization primarily interested in the welfare of telephone cooperatives.

"Our annual meeting in February attracted the largest attendance in the short history of the association. We hope to increase in size each year until every telephone co-op in the country is enjoying the advantages of belonging to the national.

"Many have asked, 'just what are the advantages of belonging to NTCA? What can they do for my co-op and what have they done the past?' Briefly, here is the answer:

"1. An excellent retirement and security program has been set up for the employees of member telephone co-ops. Insurance is through the John Hancock Mutual Life Insurance Company.

"2. Group health and accident insurance is available through NTCA. It is underwritten by Employers Mutual Casualty Insurance Co.

"3. We have a staff in Washington, D.C. to represent you before Congress and obtain answers for member inquiries.

"4. NTCA is your united cooperative organization which has fought for the (a). Reduction of capital equity requirement for cooperatives.

"(b). Extension to five years on loans before principal payments are due.

"(c). Maintaining the present two percent interest rate on REA loans.

"(d). Has appeared each year before congressional committees to ask for REA telephone loan appropriations.

"(e). Has taken a stand against private financing of REA loans.

"(f). Has supported the Humphrey-Price bill to restore REA administrators full authority over REA.

"(g). Has worked in behalf of better toll settlement agreements.

"(h). Has repeatedly advocated that cost of EAS between exchanges should be split 50-50 without any other charges.

"(i). Has worked for better operator office assistance contracts.

"(j). Worked with REA to start REA telephone borrower institutes.

"(k). Has urged the repeal of the 10 percent excise tax.

"(l). Has lobbied against any changes in the federal tax exemptions for co-ops.

"(m). Has initiated special NTCA management institutes.

"(n). Has intervened in state regulatory commission cases where a co-op was discriminated against.

"(o). Has fought REA's previous insistence that co-ops obtain letters of territorial rights from neighboring companies prior to a loan application.

"(p). Has filed objections with USITA where independently owned toll centers have not given the telephone co-ops the extra cash benefits obtained through the new toll settlement agreements with Bell.

"(q). Asked REA to set up a national job training and safety program.

"(r). Asked REA to permit telephone cooperatives (at their option) to repay previously collected equity payments from loan funds.

"(s). NTCA is working with REA and the industry to set up a compatible telephone mobile radio system with industry-wide participation.

"We have almost filled the alphabet by giving examples of the work NTCA has done.

"If you are not a member of NTCA, I urge you to join now. We are all in the same boat so let's row together!!"

Briefs -1959

On March 26, an early morning sleet storm increased the ice loading on all lines. The loading and the gusts of wind leaned over a great number of poles on the north and south lines and about 60 poles were broken on the Winnebago RTA lines.

—

A new Iowa law provides for a $100 fine or a 30-day jail sentence for any person who fails to relinquish a telephone line when another party requests the line for an emergency call.

—

Paul Nordskog, rural Forest City, has private line telephone service via radio telephone carrier over the same pair of party-line wires his neighbors use.

—

From the September *Dial Tone*: "Notices have been mailed out to each member showing the amount of deferred patronage dividends that have accrued to your account. Being the Cooperative has borrowed $1,126,647.58 from the government to build your system, these dividends are being used to pay off the Cooperative's indebtedness and for replacement of the Cooperative's system as needed. Therefore, you are gaining an ownership investment in the system in proportion to your patronage. This investment in reality is in poles, wires, switchboards, telephones, etc. Your dividend will remain invested in the Cooperative until such time as sufficient funds have been accumulated, after which the Board of Directors may retire these credits in the order in which they have been accumulated."

—

By the end of 1959, the Cooperative had paid $166,664.59 interest and principal on its REA loan. $165,342.67 is now invested in U.S. bonds for future replacements and storm fund.

1960

On March 15, 1960, 177 stockholders of the Grafton Telephone Company voted 149-20 to join the Winnebago RTA. The move was occasioned by application of age and hour laws to the Grafton company, which would have forced it to pay additional operator salaries. The assets of the Grafton company were purchased by the RTA.

Officers of the Grafton company were Wilbur Biederman, president; Kermit Twettin, vice president; Harry Diedrick, secretary; directors were Henry Hartman, William Rappath, Vern White, Eugene Wild.

The Grafton news was picked up by the *Phone Call*, published by the National Telephone Cooperative Association, which reported in September, 1960, that "about a year from now, the 275 telephone subscribers in Grafton, Iowa, will be able to talk to their neighbors nearby and in surrounding areas without interference. . . . They'll get the modern, up-to-date, reliable service with the same, or lower, rates that they pay themselves. How will this come about? What will bring Grafton into the communication pattern enjoyed by most Americans, though only about 60 percent of rural families? Grafton's Mutual Telephone Company has incorporated its facilities with the nearby Winnebago Cooperative Association.

"Winnebago's engineers and crews will completely rebuild and modernize the system, install dial phones and bury all the lines in Grafton and 72 miles of rural lines where no storm can bother them. They'll double the number of toll circuits and triple or quadruple the number of lines serving the dial service area—the 'free' area . . . and some new subscribers, who never before had phone service and couldn't get any, will now be hooked on the Winnebago line."

Sound familiar? Grafton joins the other exchanges in the Winnebago RTA. And still, outside the Winnebago area, four in ten American farms were without phone service in 1960.

In March, 1960, Glenn Bergland was re-elected president of the National Telephone Cooperative Association at its annual meeting in St. Louis.

At the meeting, Mr. Bergland pointed out that two-thirds of the money loaned since the inception of the REA telephone loan program went to commercial rather than co-op borrowers. "However," he added, "the commercial borrowers are not building two-thirds of the lines for which the money has been loaned, but, in fact, 20,000 fewer miles of lines. Cooperative borrowers have 2.6 members per mile compared to five subscribers per mile for commercial borrowers."

In testimony given to the U.S. House of Representatives' Appropriations Subcommittee, NTCA president Bergland said NTCA had surveyed all telephone cooperatives and received an 80 percent reply, and that based on this and allowing for an unknown number of new borrowers that may apply for telephone loans, REA loan funds of $50 million should be set aside for telephone co-op borrowers for fiscal 1961.

In May 1960, the Cooperative filed for a $130,000 REA loan to convert the Grafton exchange to dial in 1961.

Effective June 4, 1960, Marlo Durby, the Cooperative's central office equipment and maintenance man, was put in charge of maintenance in the Conger, Emmons, Joice, Lake Mills, Scarville and Leland exchanges. The Durbys moved from Thompson to a home they purchased in Lake Mills.

On August 15, 1960, stockholders of the Fertile (Ia.) Incorporated Telephone Company voted unanimously to sell their company to the Winnebago RTA for $31,500.00.

RTA Manager Bergland said a new building would be constructed in Fertile to house an office and switchboard equipment, and that the entire Fertile telephone system would be rewired for a conversion to dial in 1961.

The addition of Fertile's 280 subscribers made a total of 4,800 RTA subscribers.

The Fertile Incorporated Telephone Company board of directors comprised Maynard Calgaard, President; Nels Fjeld, Secretary; and Roger Humphrey, Gussie Haugen, Henry VanDenBosch, Harry Rodberg and Clark Halvorson.

Mr. Fjeld was ninety years old and had served sixty years as a director of the company, forty years as Secretary-Treasurer.

The Cooperative applied for a $159,000 REA loan to finance the rebuilding of the Fertile phone system.

On September 7, 1960, the residents of Fertile and Grafton voted 45-0 and 55-0, respectively, to grant a 25-year franchise to the Cooperative for the use of the towns' streets, alleys and public places for the new telephone systems.

In September, 1960, *Dial Tone*, Manager Bergland addressed a letter to the Fertile subscribers, which read in part:

"Dear Fertile Subscriber:

"We were most encouraged to find there were no opposing votes in the franchise election. . . . On September 8, we went through the records with the Fertile Telephone Company officials. Arrangements were made for us to collect their old bills and tolls prior to September 1 and we are working out a billing for October 1. . . . Previously your company has billed quarterly. We will bill monthly.

"We have retained the services of your present operators, Aimee Nye and Myna Smith. Mr. George Hall will continue to repair trouble on the old system until such time that we cut over to dial. . . .

"At any time you have any questions relative to our operation, we will be most happy if you will contact one of our men who will be working in your area or write our office in Thompson.

"We wish to be of service to you and render you the best possible type of telephone communication.

Very truly yours,
Glenn Bergland, Manager"

(Aimee Nye had thirty-one years of service with the Fertile company, and Myna Smith, twenty-six years.)

The Stromberg Carlson Company was the low bidder at $23,472.77 for the new dial switchboard for the Grafton exchange.

All of the rebuilt Grafton telephone system was buried, both in town and rural. The cost of the buried plant was higher than that of an open wire plant, but there was less maintenance and less trouble from lightning and storm. Part of the new Fertile system was buried, but the Cooperative used all of the Fertile overhead lines that had been built in the prior six or seven years because they represented a considerable investment which could be used.

Briefs - 1960

The 1960 property taxes of $16,361.50 paid by the Cooperative make it one of the highest property tax payers in this area. The sum amounts to $5.11 for each subscriber per year.

—

The addition of the Grafton exchange to the Winnebago RTA brings to 12 the number of exchanges in the Cooperative.

—

Telephone service for fire departments in all towns where the Winnebago RTA has an exchange became effective May 1. Benefited are the towns of Thompson, Rake, Leland, Scarville, Joice and Crystal Lake in Iowa, and Emmons and Conger in Minnesota.

—

In April, the Cooperative signed agreements with the Manly and Kensett telephone companies for free service circuits to those towns, and expects to sign with Plymouth for the same.

—

As of May, the Cooperative has negotiated and signed boundary agreements with Manly, Kensett, Northwood, St. Ansgar and Osage, and negotiations are underway for the Plymouth boundary.

—

The Cooperative notified the Grafton subscribers that those who are not stockholders in the old Grafton company will pay a $10 refundable membership fee to the RTA. For stockholders, the fee will be paid by the Grafton company. The Cooperative inventoried and appraised the entire Grafton exchange and prepared plat maps for the town and rural subscribers.

—

On May 9, the *New York Times* phoned to ask the Cooperative the name of the most recently connected REA and RTA member. It was Floyd Petersburg, rural Joice. The next day the *Times* phoned Mr. and Mrs. Petersburg and asked them questions about their telephone service and what it was like before they had electricity, how many towns they could call free, what they thought of the future of farming, etc. The phone call was reported in the *Times*.

—

In retrospect, there is some humor in the following notice in the August *Dial Tone*:

Attention, Leland subscribers! In the yellow section of your directory, please change the number of the Forest City Rendering Works to 2-411. A doctor in Forest City has been receiving many of these calls due to the wrong number.

—

Recently squirrels gnawed holes in the Emmons-Twin Lakes overhead cable, disrupting service. Evidently, at a certain time of the year squirrels have an appetite for some of the salts found in the cable's lead covering.

—

The Cooperative secured a boundary agreement with the Hanlontown Telephone Company for three free service lines between Hanlontown and Fertile; with the Clear Lake Telephone Company for a boundary between Fertile and Clear Lake; with Northwestern Bell for a boundary between

Mason City and Fertile, and with the Ventura Telephone Company for a boundary between Ventura and Fertile.

—

On September 2, Clifford Brackey discontinued RTA telephone service by the Lake Mills exchange. In 1951, Mr. Brackey was assigned RTA membership certificate No. 1.

The November *Dial Tone* reports that "Mr. Kenny Lein, our central office equipment technician, has begun installing new radio carrier equipment for additional circuits from Thompson to Scarville and Crystal Lake. It was found to be cheaper to add these circuits by use of radio carrier instead of extra telephone lines."

The lawful statement of the ownership, management and circulation of the *Dial Tone* was published in its November issue. "Owner: Winnebago RTA; Editor, managing editor and business manager: Glenn Bergland; known bondholders and other security holders: U.S. Government, Washington, D. C. Circulation: 3,067."

1961

Glenn Bergland completed two terms as president of the National Telephone Cooperative Association, effective with the annual meeting of the NTCA in Dallas, Feb. 9-11, 1961. Glenn had announced in Fall '60 that he did not wish to be considered for re-election. Attending with Manager Bergland were Cooperative board members R.E. Aukes, Archie Anderson, Arlo Carson, Elmer Seim, Gustaf Gustafson, Norman Larson and M.J. Madson.

(Also in February, 1961, REA Administrator Norman Clapp offered Manager Bergland the position of REA Assistant Administrator in Washington, D.C. Manager Bergland, after some thought, declined. Manager Bergland was again offered the same position by REA Administrator David Hamil in 1971. He also declined this offer after more thought.)

"It was a pleasure to serve the N.T.C.A.," Glenn said. "The experiences were interesting, rewarding and valuable. Met a great number of interesting and admirable people. Many of these acquaintances will be cherished for a great number of years. Appearances before various Congressional committees and many governmental agencies proved rewarding experiences."

In March 1961, Midwest Contractors of Dodge City, Kansas, was awarded the contract for over ninety miles of plowed-in buried cable for all of the Grafton exchange and Fertile town and one-third of Fertile rural, with a bid of $87,473.34, the lowest of eight, the highest being $106,927.00.

The cable to be used was a new type, suitable for underground use, with adequate protection against rodents. The trenching plow is usually pulled by two caterpillar tractors and can plow four to seven miles a day right down the middle of a graveled road, and within two or three days it is impossible to detect where the line was plowed. Take-offs are plowed or trenched directly to the houses, with the entrance brought into the basement. (For the most part, cables were not laid in roads or public right-of-ways; most were laid in private property right-of-ways.)

On March 12, 1961, the Nonnweiler E. Richardt Construction Company, Garner, was awarded the contract for new switchboard buildings in Fertile and Grafton, for $12,135, the lowest of four bids.

On May 13, 1961, members of the Kensett Telephone Company voted 186-52 to join the Winnebago RTA. The Kensett company would continue operations until September 1, the date of changeover to RTA.

In July, the Cooperative equipped its five maintenance trucks with two-way mobile radios for instant two-way communication with the home office. Before this, the maintenance men periodically phoned the office for information and assignments. A radio call to an unattended truck either automatically turned on the truck lights or blew the truck horn to summon the driver. The base transmitter, receiver stations and tower were on a hill three miles south of Thompson.

In a unanimous vote on August 16, 1961, Kensett residents gave the Cooperative a 25-year franchise to operate a telephone system in Kensett.

On August 24, Grafton was cut over to dial and on September 19, Fertile was cut over. The two towns have free service lines between them. Grafton would also have free service lines with Manly, Kensett and Plymouth. Fertile would have free service with Joice, Forest City, and Hanlontown.

On September 1, the Cooperative assumed operations of the Kensett magneto phone system, with four operators on duty to comply with federal wage and hour provisions. Conversion to dial was estimated for August 1, 1962.

More About the Operators of Old

The conversion to the dial system meant the retirement of a great number of faithful, energetic telephone operators who had served diligently for many years. The community switchboard operator, the focal point of the community, passed from the scene.

These operators were paid very modest wages for their long hours and devoted service. Their value to the community and some of the many little favors they did for the subscribers are depicted earlier in this book.

In almost all of the towns the people would have a farewell party for the retired operators.

In Kensett, the townspeople turned out to honor the two operators who between them had given seventy-seven years of service to the town and rural patrons. Melvina Arendts and Mrs. Selma Olmsted began their work as switchboard operators in 1907. To the Kensett patrons, they were known as Molly and Sel.

With the cutover of the Emmons and Conger telephone systems to dial in January, 1956, and the retirement of the two operators, a tradition ended. Mrs. Melita Fink, who operated the Conger switchboard for many years, said she would miss the 24 hour-a-day job for her and her husband, Mel Fink. The following is excerpted from an article printed at the time in the *Albert Lea Tribune*, January, 1956:

> A flap on the Conger telephone switchboard dropped into place with a buzz. Melita Fink left her baking, wiped the flour off her hands, and sat down in front of the instrument. Calling the woman on the line by her first name, Melita asked, "You want to talk to your daughter?"
>
> "Yes," replied the woman.
>
> "She just left the store," said Melita. "Wait a minute."
>
> Looking out the window, she watched a young woman walk down the street and turn in at a certain house.
>
> Leaning toward the board's mouthpiece, she finished her duty toward the subscriber.
>
> "She's just going into your house," said Melita.
>
> A dial phone can't tell you things like that.
>
> It looks like the dial system is here to stay. In many ways it is better than the old way.
>
> But did you ever ask a dial phone what the weather report says? Or the name of the new family that moved into town? Or what was on special at the town's grocery?

Pictured in the *Lake Mills Graphic* of the day was the old switchboard at Emmons with Mrs. Andrew Rasmussen, who had served as switchboard operator for 26 years. Also in the picture were Mrs. J. A. Wilson, switchboard operator, S. R. Emerson, who was manager of the Farmers Telephone Company, Emmons and Conger, before it was merged into the Winnebago RTA, and Jim Beaudy, maintenance man for the old company.

The Farmers Mutual Switchboard Company switchboard in Forest City was operated by Pearl Oswald and Nora Hanna for twenty-eight and twenty-four years, respectively. The two women were featured in the *Forest City Summit* newspaper on the occasion of the changeover to automatic dial switchboards, the Farmers Mutual Company having joined the Winnebago RTA. (The Farmers Mutual was organized about 1900 and was owned by eighteen small telephone companies at the time of consolidation with the RTA.)

On September 16, 1961, the Fertile community held a retirement tea for its operators, Mrs. Aimee Nye and Mrs. Myna Smith, and for Secretary-Treasurer Nels Fjeld and lineman George Hall. A great number of people attended the tea and extended their appreciation for the wonderful service they had received. The following excerpts from that evening's program are reprinted here because they are typical of any community that had a magneto system and then cut over to dial:

Farewell to the Operators and Our Telephone Company

By Mrs. George Hall, Fertile

Hark, the telephones are ringing and they are very new,
For we have changed from black ones to yellow, pink and blue.
And that is why we are here tonight,
To bid farewell to a familiar sight;
To a telephone office and its operators two,
Who will leave this scene for a life that's new.
Let's turn back the pages of Old Father Time
And reminisce in rhythm and rhyme.
The first switchboard found a tiny niche
In E.E. Brown's Meat Market where they all could switch.
The calls that came through were very few,
So it wasn't hard to take care of the two.
But the time soon came to change its place;
They built a new room and lifted its face.
The Browns were there as the keys fell down
To answer the calls for the old hometown.
It changed hands quite rapid as time went by
And several people thought they'd give it a try,
But at the end of a year or two
They'd announce that they were through.
So in 1929 two new operators finally arrived
With two small boys who were quite alive;
And Aimee Nye with Norman and Max
Moved in and took over and relaxed;
And Bess Purcell Rodberg moved in that day
and the Fertile switchboard was far on its way.
In '36 another change was due
And they all moved to a place that was new.
They found themselves right on Main Street,
And don't ever think that wasn't a treat.
For the public asked every question under the sun
And were sure the operators could answer each one.
By 1937 Myna Smith and Darline joined the force
And Myna was taking a telephone course.
She helped each time as she saw their need
And in 1948 become an operator indeed.
To the best of their ability they have served the town
And switched for the country folk for miles around.
No calls were to be answered after nine
But some people forgot to look at their time.
They rang at nine or eleven or two
And expected each call to go right through.

They reported new babies here and there,
How much they weighed and what color hair.
They'd ring the fire whistle strong and loud
And it would always draw quite a crowd.
The keys would drop down like stars in the sky
Every time the fire engine passed by.
They'd hear voices ask, "Where's the fire today?"
And they'd all get an answer somehow, some way.
Husbands would get lost and couldn't be found
And they'd ask the operator if they'd please look around
And if they could see him on the street,
Would they please send him home with the luncheon meat.
Some poor little mother just pulling her hair
Would ask the operators if her kids were there,
And would she please tell them to stay right by the door,
She'd be there when she got the wax on the floor.
The doctor would call that his vacation was due
And a lot of his patients were feeling blue,
But if anyone fell and broke their leg,
Send them to hospital to get it pegged.
The gas truckers thought it would be nice
If the operators would take their calls instead of their wives.
They had to set by the board anyway
And why keep their wives home all day.
They had given death messages and heard its knell
And listened to many a wedding bell.
They have seen other children as well as their own
Get their education, grow up and leave home.
For when you've spent so much of your life at one thing,
There's bound to be progress, even in a telephone ring.
At one time, they cranked every number by hand,
Then came electricity, now the dial system plan.
The telephone company needed more than just ladies' frills,
They needed a secretary and treasurer to pay the bills.
Nels Fjeld was elected with paper and pen
And after forty years his work has come to an end.
Then the lines got broken, telephones wouldn't ring
And they needed a lineman to mend everything.
So George Hall applied, got his tools and was hired.
He worked twenty-five years, they went to dial and he retired.
So we say farewell as we blot out our tears
To a telephone company that's been here fifty years;
And to the operators, who have worked so true,
And we'll accept the dial system and try something new.

A Typical Day in the Old Fertile Telephone Office

by Mrs. C. Norman Nye and
Mrs. Roger Humphrey

Time: Early one morning-

Night bell rings—operator comes to board.

Op: "They sure are starting the day early, it isn't even time to open up. Fertile—is this an emergency? We aren't open yet."

Party: "Oh, am I too early to ring? What time is it?"

Op: "It's only ten to 7:00, but I'll ring for you."

Party: "My clock says a little after 7:00. I'm sorry I called."

RING

Op: "Fertile."

Party: "Say, Aimee, I'm going away for a while this forenoon and I wondered if you'd take our orders for me?"

Op: "How long are you going to be gone?"

Party: "Oh, I should be back in a couple of hours."

Op: "Yes, I will."

RING

Op: "Fertile"

Party: "Have you seen my husband up there on the street?"

Op: "No, not just lately."

Party: "You don't see our car do you?"

Op: "No, but just a minute and I'll go and see." (Operator goes to door and looks up and down the street.) "Yes, I believe it is up at Ernie's garage."

Party: Will you ring there for me, please."

RING

Op: "Fertile"

Party: "Have you heard what time the shower is tonight?"

Op: "It's supposed to start at 7:30 but they never get started until 8."

Party: "Is it at the Lutheran Church or the schoolhouse? I guess I threw my card away and can't remember which it said."

Op: "It is in the school dining room."

RING

Op: "Fertile"

Party: "Ring my mommy."

Op: "Well, who are you?"

Party: "Dick."

Op: "Just a minute, Dick, and I'll ring her for you."

Op: "My land, I can't ring another soul, every plug is up and I can't get a line."

RING

Op: "Fertile."

Party: "What time do you have?"

Op: "It is a quarter to eleven."

Op: "I better blow the whistle so everyone knows it's dinnertime."

NOON WHISTLE

Op: "My land but it's getting dark outside. I wonder if it's going to storm. Just look at those clouds. I'm afraid we're in for it."

RING

Op: "Fertile." (Operator jumps) "My, but I got a shock. The sparks are just flying off the board. I wish people wouldn't ring when it storms."

RING

Op: "Fertile."

Party (frantically): "Call the fire department. The lightning struck a tree in our front yard and started a fire."

Op: "Who is this?"

Party: "Gussie Haugen."

(Siren blows—confusion reigns on the switchboard; everyone wants to know where the fire is, how bad it is or what is burning.)

Time: 9:30 P.M.

Op: "This has been a long day. Guess I'll lock up. It's 9:30."

RING

Op: "Fertile."

Party: "Say am I too late to call? I want to get in touch with my girl before she leaves the schoolhouse."

Op: "Well, it's after 9:30, but I'll ring for you." (Leaving board) "I'm so tired, I hope the night bell doesn't ring tonight."

* * *

The October, 1961, *Dial Tone* printed this expression of thanks from the Cooperative to the Grafton and Fertile telephone operators:

> "To the Grafton and Fertile Telephone Operators:
>
> "To Ms. Lizzie Kirchgatter and Mrs. Hazel Schultz of Grafton and Mesdames Myrna Smith and Aimee Nye of Fertile, telephone operators, our sincere gratitude and appreciation for the wonderful service they have performed for their communities in the last several years. Mrs. Kirchgatter and Mrs. Schultz have been operators at the Grafton switchboard for a great number of years. Mrs. Aimee Nye and Mrs. Myna Smith served the Fertile community for over 20 years. We also wish to acknowledge the wonderful service that the Fertile Sec-Treas, Mr. Nels Fjeld, provided his company during the last 50 years. We also wish to express our sincere thanks to Mr. George Hall, lineman for the Fertile Telephone Company, for the service he rendered them during the last 25 years. We are sure in many cases the communities will miss the services provided them during their years of retirement."

On August 22, 1962, 200 people in the Kensett area gave a reception and program honoring two of the town's best known citizens—operators Melvina Arendts, who served faithfully from 1915 to cutover date, and Mrs. Selma Olmsted, who served 1907-15 and again 1940 until cutover date.

Epilogue

by Tom Fretty:

This is meant as a fond farewell of memories dear of Molly and Sel.
We'll never forget your kindness and aid,
 and we're certain that someday for this you'll be paid.
When far away and on a trip,
 we'd hear that voice and nearly flipped—
 Molly was "Kinsett," Sel, "Hedoe".
The warmth of that answer would make our hearts glow.
Fire, funerals, snow and sleet,
 information was free and always complete.
Those voices we hear on our party line phone
 will now be replaced by dial tone.
The hole is round, our finger will get sore—
 we just can't give the name anymore.
The old switchboard, a chair and a table—
 all this we've swapped for an underground cable.
We'll miss you, ladies, and your friendly "flatter"—
 your kindness, good natured warm hearted chatter.
When calling long distance to a friend far away—
 they say there should never be a delay.
But with all these improvements, we'll miss you a mile,
 for now all we've got is that chrome-plated dial.

* * *

For the year of 1961, Manager Bergland said that 100 miles of buried cable were added to the Cooperative's system and that within the next two years, the Cooperatives expected to rebuild 100 miles of old line in the Thompson and Buffalo Center exchanges. The Cooperative had income of $236,549.37 and expenses of $195,666.26, leaving $40,883.11 to be applied on the REA loan.

It is interesting that back then, of the 4,069 phones serviced, there were just fourteen rural private party lines.

Briefs - 1961

On April 24, the Freeborn County maintenance crew cut a 52-wire buried cable south of Conger, putting every telephone south, east and west of Conger out of service until the Cooperative was able to splice the cable the following day.

—

The Fertile and Grafton fire departments have been equipped with a fire conference circuit, a new adaptation in fire alarm circuits not previously available to the Co-op's other exchanges. When the fire number is dialed, it will automatically ring ten volunteer firemen's telephones. If their phone(s) happen to be busy, they will get a special warning sound indicating both parties should hang up as a fire call is coming in. All vacant lines will receive the ring and will be able to converse even before the busy lines are cleared.

—

Two copper wires have been installed between Fertile and Forest City over which two channels of radio carrier will be added to provide three free service lines between the two towns.

—

Regarding old poles and wire in the August *Dial Tone*: Many Grafton and Fertile subscribers have asked us our policy concerning the old poles that will not be used after dial cutover. Our policy is as follows: (1) Our crews will go through the entire system and remove for our own use all creosoted poles. (2) After we have removed the poles we desire to keep, a letter will be sent to all subscribers granting land owners 60 days to remove all other poles and wire adjacent to their farms. (3) After expiration of this period, we in turn will give these poles and wires to any others who may be interested, as we do want them removed.

—

Manager Glenn Bergland and technician Kenneth Lein attended a week's telephone engineering school in Minneapolis Nov. 27 through Dec. 1.

"Molly and Sel," beloved operators of the Kensett switchboard. Melvina Arendts was operator from 1915 to August 28, 1962, the conversion date to RTA dial. Mrs. Selma Olmsted was operator from 1907 to 1915 and again from 1940 to August 28, 1962.

1962

At the Cooperative's annual meeting in March, 1962, the board of directors re-elected the following officers: R.E. Aukes, President; M. J. Madson, Vice President; Gustaf Gustafson, Sec-Treas.

Contracts were let in May, 1962, for $26,979.65 to Stromberg Carlson Company for a new dial switchboard for Kensett, and to Midwest Utilities Company, Prior Lake, Minn., for $74,407.67 to bury the Kensett exchange telephone lines.

In separate elections, August-September, 1962, the seven companies that jointly owned the Woden Telephone exchange voted to join the Winnebago RTA. Elections were as follows:

August 10, Woden Farmers Telephone Company, Edward Cink, President, Julius Ostercamp, Secretary; South Woden Cooperative Telephone Company, Thomas Hinders, President, Harley Lumberg, Secretary.

August 17, Tri-county Cooperative Telephone Company, Herman Limberg, President, Jim Blakely, Secretary.

September 4, Lake George Telephone Company, Fred Hugo, President, Francis Harrington, Sec-Treas, James Buns, Switchboard Director.

September 6, Hollandale Mutual Telephone Company, Ronald Barnes, President, T.J. Nodland, Sec-Treas, Walter Logemann, Switchboard Director.

September 27, Woden Cooperative Telephone Association, H.B. Eisebrand, President, William J. Wood, Secretary.

On November 1, the residents of the town of Woden voted 38-2 to grant a 25-year franchise to the RTA.

Following the above elections, the Winnebago RTA immediately applied for a $160,000 REA loan to rebuild the Woden exchange and part of the Crystal Lake exchange.

On August 28, 1962, Kensett was cut over to dial. O.K. Storre, who was one of those instrumental in promoting RTA in the community, made the first long distance call, to his friend, Art Sineby, at the Pick-Congress hotel in Chicago.

On October 1, the Britt-Crystal Lake Telephone Company's board of directors—Ed Korsa, president, Paul Madson, secretary—sold their lines serving 66 subscribers to the Winnebago RTA, and sold the lines serving 35 subscribers to the Bell Telephone Company out of Britt. The Britt-Crystal Lake company was then purchasing switching service from the Winnebago RTA out of Crystal Lake and out of the Bell system at Britt and maintaining their own lines. They found, however, that the cost of rebuilding and repair was excessive and that it was advisable to turn over their facilities to the respective companies, i.e., Bell and the Winnebago RTA.

For 1962, the Cooperative had income of $284,865.04, and expenses of $236,975, leaving $47,889.62 to be applied on the REA loan. Included in the expenses were depreciation of $65,093.90 and interest on the REA loan of $23,602.42.

Briefs - 1962

The Cooperative notified its subscribers that it "will not allow or tolerate any foreign attachments on our telephone lines such as extensions, telephone bells, etc. Our employees finding any of these will remove them. The guilty party will be billed for past services plus a penalty and loss of instrument. Failure to comply means disconnection of telephone service."

(In 1977 the FCC changed this policy.)

—

Des Moines Sunday Register, May 4: "Youngsters yakkety-yak just as much in the smaller towns as they do in the larger city.

"At least they tell us that up in Winnebago County, the independent telephone company of Thompson (pop. 698) has a special section for teenagers' telephone numbers in its new directory." (The independent telephone company is the Winnebago RTA.)

—

The November *Dial Tone* had an article, "A Day with Serviceman Ken Lein." The day included installing, removing and testing telephones, correcting switchboard trouble in the central office building in Thompson, checking the Emmons switchboard, taking telephone usage readings, repairing and replacing broken insulators south of Conger and repairing the Conger switchboard.

1963

R. E. Aukes, M. J. Madson and Gustaf Gustafson were re-elected President, Vice President and Sec.-Treas., respectively, of the RTA board of directors following the annual meeting on March 27, 1963.

RTA Escapes Rate Regulation

In 1963, the telephone and electric rates in Iowa were not regulated by the state, but supposedly were regulated by the town and city councils in the areas served. Most of the councils lacked both the expertise and financial resources to fairly set rates and consequently there were many law suits. To remedy this condition, the power companies and large telephone companies sought to establish a single state regulatory commission under which their rates would be set.

At the time, Manager Bergland was on the regulatory committee of the Iowa Telephone Association and the Iowa Association of Electric Cooperatives, which had not yet taken a position of whether or not they were in favor of being subject to the provisions of a state rate commission. The committee researched several states and learned that electric and telephone co-ops not subject to commission regulation found themselves in a kind of no-man's land as to territory and prey for the larger regulated companies. For that reason, they conceived the theory that the co-ops would be regulated as to territory and service, but be exempt from the rate regulation provisions of the proposed act. They also extended this provision to small commercial telephone companies with fewer than 2,000 stations.

Manager Bergland took the position that since telephone cooperatives operated at cost and were thereby self-regulated by their members, state rate regulation of cooperatives and small companies would be both unnecessary and a bureaucratic expense.

On the last day of the House debate on the bill, Rep. Henry C. Nelson of Forest City introduced an amendment sponsored by Manager Bergland that provided that the Iowa telephone cooperatives could upon written application be exempted from regulation by the Federal Communications Commission so long as the cooperatives were state regulated. Previously, the FCC had threatened the cooperatives with federal regulation when a system had lines on both sides of a state line and were not regulated by the outside state. Hence, to have an alternative in case of future federal difficulties, the cooperatives could choose which—federal or state—would have regulatory authority over them.

All large telephone companies were now rate regulated, as well as about five smaller independent companies. The latter included Western Iowa Telephone, which was headquartered at West Bend, Iowa. Because Western Iowa Telephone was planning a future rebuild of its system, it had saved up a cash reserve as equity for a commercial loan, the REA loans program being unavailable to the company. The state regulatory commission looked at the cash reserve and, oblivious to its purpose, not only put a lid on Western Iowa's rates, but decreed they reduce their monthly rate to a ridiculous 50 cents per month.

"This was asinine," Manager Bergland says. "As a result, within two years Western Iowa was almost broke.

"Because of this, our state committee sought to amend the law to exempt from rate regulation any company with fewer than 25,000 subscribers.

"We solicited the assistance of Lt. Governor Terry Branstad, who was also president of the Iowa Senate, and in the late '70's we were able to obtain legislation to exempt independent telephone companies with fewer than 15,000 subscribers. 'Independents' are those companies not part of the Bell system."

Considerable time and effort in drafting the language of the Utilities Commission Bill was spent by Manager Bergland and the regulatory committee of the Iowa Telephone Association.

The Cooperative filed its rules and regulations, totalling 66 pages, together with maps outlining exchange boundaries, with the new state commission.

The Cooperative contributes a share of the funds required to operate the commission.

In July, 1963, the Cooperative was approved by the REA for a $278,000 loan to rebuild the Woden and Crystal Lake exchanges. The sum brought to $1,793,000 the Cooperative's debt to the REA.

Also in July, 1963, the Cooperative awarded a $45,867.01 contract to Stromberg Carlson Co. for two new dial switchboards for Crystal Lake and Woden, and a $92,201.80 contract to Hanson and Aune Construction of Ironton, Minn., to bury 106.2 miles of telephone lines in the Crystal

Lake and Woden exchanges. Bids on the latter were close, the second and third low bids being $94,142.91 and $97,240.64.

In 1963 the Cooperative paid property taxes of $24,443.12, assessed on its property in the counties of Worth, Winnebago, Mitchell, Kossuth, Hancock and Cerro Gordo.

The Cooperative had revenues in 1963 of $300,270.87 and expenses of $248,425.94, leaving $51,844.93 to apply on its REA loan. The expenses included $26,196.63 loan interest and $71,123.94 reserved for depreciation.

Briefs - 1963

Direct distance dialing is getting closer. Some of the RTA's exchanges are already equipped; others need revision. To coordinate RTA exchanges with the national direct dialing network, the telephone numbers in the next directory will show seven numbers.

—

One thousand fifty-five silver dollars were given to members and spouses who attended the RTA annual meeting and registered before 10:30 a.m.

—

A new fire reporting telephone circuit has been placed in service for the Twin Lakes fire department. The fire number is 297-2000; one or more of 10 firemen will answer.

—

More than 668 people registered at the Cooperative's booth at the Winnebago 4-H and FFA fair.

—

Twelve cases of mole gnawing were reported this summer. Moles have small teeth, which are able to penetrate the steel strands protecting the interior of the cable, puncturing the insulation and causing water to leak in.

—

Cooperative board members meet one day each month and are paid $5.00 per meeting. The payment includes transportation.

1964

The Cooperative's annual report included a very low $576.98 reserve for bad debts, which included several subscriber bankruptcies and transfer of unpaid dues from consolidated companies.

The Cooperative held its 1964 annual meeting in March, after which the following officers were re-elected by the board of directors: R. E. Aukes, President; M. J. Madson, Vice President; Gustaf Gustafson, Sec.-Treas.

On April 2, 1964, the Woden exchange was cut over to dial, with the first call make by Imko Huisinga, who was president of the old Woden Switchboard Company.

On April 9, the Crystal Lake exchange was cut over to dial, with the first call made by Woden's mayor, Frank Koenecke.

Briefs - 1964

Worth repeating, from the January *Dial Tone*:

Are you guilty of . . .

*OBSCENE LANGUAGE. A mark of ignorance.

*TALKING UNTIL CUT OFF. Then calling again immediately.

*EAVESDROPPING. Why is THEIR business yours?

*IMPATIENCE. Ask for the line courteously.

*CHILDREN PLAYING WITH PHONE. You may miss an important call and others are denied use of line, also.

—

Our Annual Meeting Day March 24 in the Forest City Civic Auditorium, the Forest City Congregational Church will be selling noonday luncheons in their church basement and the Methodist ladies will serve luncheon in the Forest City fire hall.

—

Eleven hundred silver dollars were given as attendance prizes to members and spouses at the annual meeting.

—

From the May *Dial Tone*:

Listening Buttons

In our first exchanges we installed on our telephones what we called a press-to-talk button, and to some it has been misused as a listening button. We have received numerous complaints and names of people who leave their receivers off the hook so as not to miss any neighbor's call. Actually, telephones are to be considered private and if you do not receive a ring on your phone, it means that the telephone call is a private matter between two other parties and was not intended for you.

—

"The seven ages of women are her right age and six wild guesses. Even back in the stone age when they wrote down their age they were chiseling." —Red Blanchard.

—

Lightning Labor Day night shorted several Twin Lakes lines and entered the underground cable south of Buffalo Center. It is not unusual for a lightning storm to disrupt service to five percent of the Cooperative's subscribers.

—

In October, Marlin Senne became Outside Telephone Superintendent to fill the vacancy left by the resignation of Don Jeffries, who moved to Bend, Oregon. Marlin's immediate supervisor will be Kenneth Lein, now Operations Engineer.

—

Edwin Smith, Smith Electric Plumbing and Heating, Thompson, now has an "electronic secretary," a phone answering machine, one of the extra services offered by the Winnebago RTA.

—

Roger Urbatch, 15, who broke his leg during football practice, has two-way communication between his home in Kensett and the Northwood high school, via the Winnebago RTA's special school-home intercom.

—

The Cooperative encourages its employees to enroll in Dale Carnegie classes and reimburses them upon their completion of the course.

1965

The Cooperative held its 1965 annual meeting in the month of March, as per past practice.

Manager Bergland reported that "your Cooperative staff is preparing a comprehensive engineering plan for an ideal private line telephone system . . . a detailed revenue and expense forecast for the next 10 years . . . as basic information for future decisions."

Following the meeting, the Cooperative's board of directors re-elected R. E. Aukes, President; M. J. Madson, Vice-president, and Gustaf Gustafson, Sec.-Treas.

Moving overhead lines and buried cable where there was new road grading was a continuous expense for the Cooperative. An example of this is the work that was done during the summer of 1965. New road grades required moving eleven miles of underground cable and five miles of pole line, some lines having eighteen wires. In addition, on road grades adjacent to underground cable, many laterals and pedestals had to be moved back and lowered.

A pedestal is a box wherein are connected subscriber lines with lines from the buried cable, and where lines are spliced. In long circuits, pedestals may contain induction loading coils which are inserted into the line to bring the voltage and amperage back into phase to eliminate distortion. In such cases, the pedestals are positioned not more than 4,500 feet apart.

Thailand

In August, 1965, Manager Bergland was selected by the U. S. Agency for International Development—AID—and the National Rural Electric Cooperative Association for a three-months assignment in Thailand.

In Thailand, Mr. Bergland was responsible for and directed the preliminary planning and engineering of the electrification of 150 villages, a job for which he was eminently qualified by his record electrification of rural Winnebago county and parts of adjacent counties. Manager Bergland's increasing national prominence in the fields of rural electrification and telephony, and his leadership in state and national organizations representing the two industries, were the principal factors in his being selected for the Thailand assignment.

Manager Bergland, accompanied by Mrs. Bergland and an assistant, Alex Porter, arrived on October 27, 1965, in Bangkok, capitol of Thailand.

Thailand is in Southeast Asia, adjacent to the west from Cambodia and to the south from Laos. It is about four-fifths the size of Texas, with hot and humid weather most of the year. Northeast Thailand, where Manager Bergland worked, has dry and dusty weather immediately following the annual summer monsoon season, which ends in October. Because of the government's hydroelectric projects, in part financed by UN, AID and other foreign funds, there was adequate power for establishing rural electrification where none then existed.

Below is a letter home from Manager Bergland, which was printed in the December 1965 *Dial Tone*:

"As one steps out of the door of one of the finest hotels in Bangkok, he is confronted with the stagnant odor of the klongs (canals), where men fish, children swim and women wash their clothes. As you walk across the street, you may dodge on the one hand a hawker carrying an entire restaurant balanced on a carrying pole across his shoulders, and on the other hand big buses, Jaguars and motor bikes.

"On the sidewalk, one makes his way between an ancient woman dressed in the old paisan, with a lion-manned haircut and betel nut stained mouth and teeth, nude children, and then be confronted with a young Thai girl, trim and shapely, modern as tomorrow, carrying her school books.

"Every male during his lifetime feels an obligation to serve as a Buddhist monk for at least three months. During this tenure, he is without 'worry.' His food is obtained from religious donors. Some women get up a five o'clock in the morning to begin cooking food for the Buddhist monks, who come by in saffron robes carrying their brass alms bowls. Every morning I see them stopping at various food vendors and homes, where handouts are ready and are given with reverence."

The greater portion of the electrification project was financed by the Thai government. Preliminary technical assistance was given by Manager Bergland and staff, who prepared preliminary engineering specifications, and expense and income projections. In just ten days, thirty-eight Thais helping Manager Bergland signed up 8,141 members for the new Thai electric cooperative in the Nong Hong area in Northeast Thailand near Laos.

Other work done by Manager Bergland during his three-months stay in Thailand included meeting with the Thai officials in Bangkok. "I had a desk in the office of the Director General of National Development, Khun Skun, and also a desk in the U. S. Agency of International Development," Glenn recalls. "I would shuttle between the two offices."

The Conversion From Multi-Party to Private Party Service

By 1964, the price of copper shielded polyethylene cable was reduced, making it feasible to consider converting the Winnebago RTA to all private telephone service with storm proof buried cable.

The cable consisted of copper telephone wires covered by a polyethylene sheath, which in turn was sheathed in copper, and the copper covered with a second sheath of polyethylene. The number of wires in a cable could range from a single pair to hundreds of pairs, depending upon the number of phone installations to be served. Each private line connection required its own pair of wires. A gopher might gnaw through the outside sheath of polyethylene, but be thwarted by the copper, which kept the inner polyethylene sheath intact, preventing moisture from entering and corroding the wires by electrolysis, which would cause problems and noisy lines.

The Board of Directors approved a private line feasibility study to determine the cost of conversion for each exchange. Most of the study was done by Staff Assistant Ken Lein, aided by REA engineers.

Going all private-party rural was a controversial issue with the large independents. They were inclined to believe that conversion was not practical because subscriber rates would not support the cost.

"We at the Winnebago RTA believed otherwise," Manager Bergland said. "In a year-long study, we projected revenues and expenses over a number of years and calculated a cost of $1,870,127 to convert all our exchanges to one-party lines, with a subscriber rate between six and seven dollars per month."

The study also included projected costs on conversion to four-party aerial lines versus a one-party buried system, but there would by very little salvage value if the system subsequently was changed to one-party.

The four-party buried service in the towns and the rural seven-ten party aerial wire service (most of which was installed in the '50's, only 10 to 15 years before) would be dismantled and replaced with direct buried cables.

In September, 1965, Manager Bergland reported the study findings to the Board of Directors, and later to the Co-op membership.

"Many problems arise as a telephone system grows and usage increases," Manager Bergland said. "The Cooperative is faced with huge investments to keep pace. We need new switchboards, new cables and extra circuits. When the costs were compiled, we found we would spend several hundred thousand dollars and still have only party line service.

"Our study calculated a cost of $1,870,127 to convert the following areas to one-party line service: Conger, Emmons, Twin Lakes, Joice, Leland, Scarville, Thompson and Rake; and Lake Mills, Forest City and Buffalo Center rural.

"The Cooperative and its board of directors, however, will not institute a conversion of any exchange unless substantially more than a majority of its members give their written approval," emphasized Manager Bergland.

The Cooperative then printed in the *Dial Tone* a simple mail-in ballot for the subscribers to indicate yes or no for "private telephone service at a higher rate."

The Co-op also made a tentative application to the REA for a loan of $1,635,000, which sum, with the addition of $235,127 from the Co-op's general fund, would total the $1,870,127 required to make the conversion.

1966 Storm

However, before an answer on the loan application was received from REA and before the Cooperative could compile its memberships pros and cons over conversion to private telephone service, Mother Nature interceded.

On March 22 and 23, 1966, the worst sleet and snow storm in decades struck the Winnebago area.

Approximately three to four inches of crusted snow and ice clung to the east-west telephone lines. The weight and the wind laid flat 160 miles of poles and wires. Three thousand phones were without service. The roads were impassable with the deep snow.

The Cooperative sought outside help to restore service and crews from northern Minnesota, Wisconsin, southern Missouri and southern Illinois came to the rescue. Northwestern Bell brought in fifty-two men from Missouri to repair their toll lines.

Manager Bergland was pleased and grateful to his friends for the response by outside phone companies to his calls for help.

"Phone companies never hire more help than they normally need," he said. "Yet they were willing to spare one or more crews to help out."

Some crews could only stay a few days, some longer. The crew that came the farthest was from eastern Illinois. Latest to arrive were two crews from the Red River flood area of western Minnesota.

These telephone lines, bereft of further support from the splintered pole on the right, lead to nowhere. There were 160 miles of downed lines in the great ice storm of March, 1966, the landmark storm. It was the controlling factor in the Cooperative's decision to construct a storm-proof buried system.

Long distance service became available through a temporary microwave system installed by the Bell company.

The outside crews were furnished by the following companies: Grand River Mutual Telephone, Princeton, Mo.; Northeast Telephone Co.,

A bowed pole midst sunshine and shadow.

To repair the damage, where does one start?

Down and out, but not alone: See others in distance.

Crews came from afar to help restore service.

Greene City, Mo.; Chariton Valley Telephone Co-op, Bucklin, Mo.; Eastern Illinois Telephone Co., Rantoul, Ill.; McDonough Telephone Co-op, Colchester, ILL.; Adams Telephone Co-op, Camp Point, Ill.; Vernon Telephone Co-op, Westby, Wis.; Chequamegon Telephone Co-op, Cable, Wis.; Chilbardun Telephone Co-op, Dallas, Wis.; Evans Construction Co., Early, Ia., and the Champaign Ill. Telephone Company.

It can be safely assumed that the fine reputation and wide acquaintance of Manager Bergland throughout the national telephone associations was a factor in the quick response of helping hands to the Cooperative.

This used-to-be horizontal crossbar took a downward slant.

Because of the tremendous amount of work required to repair the storm's damage, it was not until April 10—almost three weeks after the storm—before the last phone connection was restored.

Although the storm interrupted the conversion decision process, it was a major factor in the decision. After temporary repairs were made, the Cooperative would have to rebuild the damaged lines at great cost and still pay off the REA loan. Wouldn't it be more prudent economically to put the extra expense into the cost effective polyethylene cable in buried system, with an increase in subscriber rates, than to continue with a storm-prone aerial system?

The Cooperative's 1966 annual meeting, scheduled for March 23, the morning following the storm, was canceled because of snowed-in members. It was re-scheduled for April 26, at which meeting the members were further informed on private line phone service.

Board member M. J. Madson canvassed subscribers in each of the exchanges to collect ballots from those who had not responded by mail. The Cooperative found enthusiastic approval for one-party service.

Kenneth Lein of the Cooperative's engineering staff, began staking for new buried lines in the Twin Lakes area; it was hoped that all the field engineering could be completed by November 1, 1966.

In late May, 1966, the Cooperative received REA approval for a $1,160,000 loan.

In September, subscribers in the Conger, Twin Lakes and Emmons exchanges signed up ninety percent in favor of private line service, exceeding the seventy-five percent requirement of the Minnesota Public Service Commission.

On January 17, 1967, the Cooperative received contractors' bids for the burying of telephone lines in the Emmons, Twin Lakes and Conger exchanges. The five bids were in a very close range, the lowest being $323,000 and the highest, $337,000. Still, the bids were in excess of the Cooperative's budget.

"We'll Use Our Own Crews"

"Through our studies and supplier materials lists, we felt that with our own construction crews we could save as much as $40,000." Manager Bergland said at the time.

Consequently, at their January (1967) meeting, the Cooperative's board of directors, upon recommendation of Manager Bergland, rejected all bids and gave Manager Bergland authority to set up a construction crew for the Co-op, to buy cable and all necessary equipment, and hire enough extra manpower to accomplish the project.

Complete plow train for burying cable: Cable plow pulled by an HD 20 Crawler tractor and, in front, an M-4 Army Tank Retriever assisting. (1967)

Wayne Roberts, a construction superintendent with many years experience, was hired as a construction superintendent for the project. (Wayne performed superbly for the Cooperative then and until his retirement in the summer of 1989.)

Wayne located an HD-20 Allis Chalmers Crawler tractor that had been rebuilt by the Minnesota School of Mines. It was purchased for $6,000, as was a used 16,000-lb. cable plow in Appleton, Wis. Both are in use today (1990) and are worth considerably more than their original cost. The Cooperative also purchased an Army surplus M4 tank retriever for auxiliary power ahead of the HD-20.

During the summer of 1967, 143 miles of cable was buried in the Twin Lakes, Conger and Emmons exchange areas. The cable crews were organized into teams: right-of-way permission, cable plowing; trenching, setting pedestals and cable splicing; back fill, fence repair and cleanup crew; combination pipe push and wire cable drop crew; a follow-up cable splicer; and settlement of claims for crop damage.

The regular Co-op personnel were supplemented by college students in the summer. In the fall and winter months the Cooperative's crews would finish the work and connect cables into the switchboards and make changeovers.

A crawler tractor, unseen on the left, is winching this RTA cable plow through a water-filled ditch to bury cable beneath the bottom of the ditch.

On February 4, 1968, Conger was "cut over" to all private-line telephone service, and Twin Lakes and Emmons followed on March 16 and April 21, respectively.

During the summer of 1968, 300 miles of cable were buried, a tremendous task, in the exchanges of Joice, Hanlontown, Leland, Rake and Scarville, subscribers of which had nearly unanimously voted for private service. Cut over to private service was late Fall 1968 for these towns, with the exception of Joice, which was cut over in 1969.

And so it went the Cooperative's subscribers voting eighty-five to ninety percent and higher for private phones.

Thompson and Lake Mills rural were cut over in 1969; Crystal Lake and Kensett in 1971; Fertile in 1972.

All the Grafton lines were buried in 1961, but conversion to all private did not occur until 1975. The Buffalo Center rural lines were buried in 1973, with conversion to private in 1978. Woden was converted to all private in March, 1975. Forest City rural was converted to private line service in 1979.

By doing its own engineering, construction, connections and switchboard installations, the Cooperative saved thousands of dollars. Savings examples:

For Hanlontown, Rake, Scarville, Leland and Joice, $98,810; for Emmons, Conger and Twin Lakes, $73,253; for Kensett and Crystal Lake, $32,579; for Thompson and Lake Mills rural, $58,619; for Buffalo Center rural, $35,000.

Through 1988, the Cooperative has saved a total of $600,000 by using its own crews. The reader will keep in mind that the Cooperative by its own in-house educational courses and its sponsorship of correspondence courses, and by its on-the-job training programs, converted laymen who knew very little about telephony to expert telephone technicians who have kept pace with the continuing improvements in the rapidly evolving telephone industry.

Briefs - 1965

Insulators were shot off poles northeast of Joice on both February 1 and 7, and earlier on poles south of Joice. Is the culprit teenager or adult, man or child, with a new gun?

—

Fourteen RTA employees are enrolled in a telephone technician's correspondence course, the books and fees for which are paid by the Cooperative, which also corrects the test papers.

—

Alas, no silver dollar attendance prizes at the annual meeting. In their place, were "crisp new dollar bills". The silver "cartwheels" of yesteryears are now fond memories for the members.

—

A fault in the 102-wire telephone cable running underwater in State Line Lake east of Emmons was found to be twenty feet from shore under four feet of water. To test the cable, maintenance men dug through three feet of frost on the east and west sides of the lake. Service was restored to 150 area telephones after new temporary cable was laid around the lake, splicing out 1,600 feet of old cable in the lake. After the weather clears, the lake cable fault spot will be sandbagged and pumped dry and permanent repairs made to the cable.

—

COMSAT, the world's first commercial communications satellite service, expects to begin service next year, and will televise via satellite the Olympic games in Japan this October.

—

Elizabeth Ryg, Thompson, confined to her bed by polio, is able to keep up with her school work via RTA radio-telephone communication with her eighth-grade schoolroom.

—

Dennis Miller was promoted to Work Order, Materials Control and Record Clerk, replacing Stanton Larson, who resigned from the Cooperative to accept a position with the REA in Owatonna, Minn.

—

Touch-tone telephone equipment for transmitting data to the U.S. Dept. of Agriculture processing center in Kansas City was installed in the Winnebago ASCS office in Thompson by Marlin Senne and Foster Hartman of the Cooperative.

Alan Swenson of the co-op and farmer are calculating amount of crop damage done when telephone cable was buried.

1966

On February 3, 1966, the Cooperative's board appointed Vernon Jacobson, 37, rural Rake, to serve on the board of directors. Vernon replaced Archie Anderson, 67, who passed away in February. Mr. Anderson served on the board from 1950 until his passing.

Because of the severe storm of March 22-23, the Cooperative's annual meeting was held April 26, 1966. Manager Bergland reported that 1965 revenue was $344,855.99 and expenses, $283,474.55, leaving $61,411.44 for application on the REA debt. Expenses included a reserve for depreciation of $79,713.80 and interest on the REA loan of $29,360.71.

Following the meeting, the board of directors re-elected R. E. Aukes, President, and M. J. Madson, Vice President; and elected board member Arlo B. Carson to the office of Sec.-Treas. Mr. Carson replaced M. J. Madson in that position. Mr. Madson resigned from the REA/RTA board because he changed his residence and moved off the Cooperative's lines. He began board service with the Winnebago REC in 1947. Mr. Madson continued to work for the Cooperative in canvassing and signing subscribers for private-line service, selling directory advertising, soliciting easements for right-of-way for the new buried lines, and settling any damages, for destruction of crops during the plowing "in" of the cable. The crews did excellent cleanup work and there were minimal small claims; crop damage, of course, was unavoidable.

Briefs - 1966

After the severe storm March 22 and 23, limited long distance telephone service was restored to the towns of Thompson, Buffalo Center and Rake March 29, by the Northwestern Bell Telephone Co. by a temporary microwave installation.

East of Thompson, the telephone messages leave the wires and "flutter through the air" for five miles before returning to the wires of the Bell system, at the highway 9-169 junction seven miles east of town.

A portable micro-wave station has been parked in the Harm Vrieze farmyard two miles east of Thompson. A carrier cable, starting in a box on the telephone pole by the White House Cafe, at the east edge of town, connects circuits from the three towns with the portable unit. The messages are sent on by micro-wave, picked up again by the station at the seven-mile junction and connected back to the wire line.

Bell technicians tried several locations in searching for a spot from which a micro-wave station in Thompson could transmit to a decoding unit by their lines along highway 69. First try was from the elevator across country to the North Prairie church corner. That didn't work so another try was made from the east edge of town to the seven-mile corner; here the Vrieze hill was in the way, so the engineers moved the station to the Vrieze farmstead . . . thus rebuilding five miles of tangled line could wait a few days more.

—

Several different shades of feeling were expressed in the responses the Cooperative received to its questionnaire concerning private line telephone service.

"I will take my phone out rather than pay six or seven dollars per month. $3.36 is far more than our use of the line."

"We are definitely in favor of private lines . . . . The irritations which arise from sharing a line with nine other families are many, and the extra cash involved would be the best anyone could spend."

"We had a very dear lady who passed away some time ago on our line who was almost blind. She used the telephone every day for hours visiting with her children, grandchildren and friends. She did this for many years. A private line would have been a blessing for her. The telephone was the biggest enjoyment the last years of her life, yet I know it bothered her children and grandchildren to have her on the line so much."

"I am much in favor of private telephones and would be willing to pay a reasonable increase in dues for such a service."

—

After the March 22 sleet storm, the Cooperative enclosed slips with its April billing to the members on which they could claim credit for the days their phone was out of service. Here is a sampling of member responses:

"Thanks for including this slip, but don't know how anyone could ask for credit because of the wonderful service which the Company gave us."

"About credit for the telephone while it wasn't working during the storm, you can just forget it, because I know it wasn't your fault and the telephone company did a splendid job of getting it working as soon as it was possible. In my business I appreciate your kindness and thoughtfulness. Thank you very much."

"I feel I should receive credit for nine days in March. Ours is still out of order and I would like another such bill as I received this time so I can fill in the rest of the days I feel I should take off our bill. Thank you.

"We're not taking any off our telephone bill. Rather leave the money for repairs. We think you did a wonderful job after the storm."

"We still do not have telephone service so really don't know how may days we will receive credit for."

—

John Julseth, 13, son of Alvin Julseth, four miles west of Kensett, is confined to his home with rheumatoid arthritis, but communicates with his high school freshman classroom via pair of a private communications lines installed by the Cooperative and Northwestern Bell. The lines run over ten miles of buried telephone cable from the farm through Kensett to Northwood High School. School Superintendent Francis Peterson requested the service.

—

Due to cost factors, the Cooperative, in its conversion to buried cable, will re-use the two-wire overhead lines leading from the buried cable to the subscriber farms. If the subscriber wants the lead-in lines buried during construction, they can be buried at the rate of twenty dollars for the first 100 feet and ten cents per foot thereafter.

—

Cooperative engineer Kenneth Lein has begun staking for buried lines in the Twin Lakes area.

—

Keeping your telephone clean is no problem. Just go over it with a slightly damp cloth. No need to worry about getting a phone antiseptically clean. Studies have shown that phones present no hazard to health.

—

Cooperative employee Darrell Pierce was presented a graduation diploma by Manager Bergland for successfully completing his basic electricity correspondence course, one of the many courses of study made available by the Cooperative to its employees on a voluntary basis for the purpose of self-improvement.

1967

In January, 1967, Manager Bergland announced the following personnel assignments, effective with the new year: Norman Risius was appointed maintenance supervisor of the eastern district, assisted by Darrell Pierce. The district included Lake Mills and points east, and Twin Lakes, Emmons and Conger, Minn. The two men were headquartered at Lake Mills, as opposed to the expense of driving each day from Thompson. Foster Hartman was designated maintenance supervisor for the western district of the Cooperative, assisted by Colonel Piersol. Marlin Senne, outside plant superintendent, would oversee the two districts and construction and maintenance and report to Ken Lein.

The Cooperative's board of directors at their January 30, 1967, meeting appointed Victor D. Knutson, 38, to the board, to fill the vacancy created by the resignation of M. J. Madson. Mr. Knutson lived on a 360 acre farm three miles south and east of Emmons in Silver Lake township.

Elmer J. Seim, 68, passed away in March, 1967. Mr. Seim was a member of the Cooperative's board of directors, serving since 1963.

Devere Steffensen of Newton township was appointed to the board to fill the vacancy created by the death of Mr. Seim.

The Cooperative's 1967 annual meeting was on March 22, at the Forest City Civic Auditorium. Manager Bergland reported revenues for 1966 of $364,862.43 and expenses of $325,863.75, leaving $38,998.68 for application of the REA loan. The expenses included a reserve for depreciation of $79,768.81 and REA loan interest of $28,423.07.

On May 3, 1967, the Fertile exchange was cut over to automatic identification equipment (ANI) for direct distance dialing (DDD) for long distance calls.

The transistorized ANI equipment transmits the number from which the call is being made and the number called to an automatic ticketing machine in Mason City by means of a unique tone for each digit. At Mason City, the information is automatically recorded for billing. This precluded the need for a toll center operator. (This procedure has since been superseded by the Cooperative's own ticketers built into its new switchboard.)

ANI equipment was installed in the Crystal Lake exchange for cut over June 7, 1967, Lake Mills rural on May 17, and Kensett and Grafton followed in late June and early July.

On June 6, 1967, the members of the Hanlontown Telephone Company voted 88-9 to join the Winnebago RTA. Charles Brunsvold, Lawrence Gesme and Jack Larson were president, vice-president and secretary, respectively, of the company. Other board members were Merwyn Kaasa, Ernest Baumann, Ralph Jennings, Orville Elvesether, Fred Faler and James Gilbertson.

The RTA now comprised 16 exchange areas: Woden, Rake, Thompson, Crystal Lake, Scarville, Leland, Joice, Fertile, Kensett, Grafton, Emmons, Conger, Lake Mills rural, Buffalo Center rural, Forest City rural, and Hanlontown. Twin Lakes got its own switchboard in early 1968 and became the seventeenth RTA exchange.

Briefs - 1967

The Iowa legislature convened on January 9. Manager Bergland is chairman of the legislative committee of the Rural Iowa Telephone Association, the purpose of which is to watch all legislation relating to commission jurisdiction, highway permits, taxation, etc., of the small Iowa telephone companies.

—

In January the Cooperative purchased from Northwestern Bell the toll lines in the Twin Lakes and Emmons exchanges, from which the Cooperative will receive additional revenue.

—

Stromberg Carlson Company will supply new switchboards for the Conger and Emmons exchanges. The replaced Conger and Emmons boards will be rewired by the RTA personnel and installed in the Twin Lakes exchange.

—

Foster Hartman and Norman Risius attended a two-week dial switchboard school in Sioux Falls, conducted by Stromberg Carlson Company.

—

On October 23, the electorate of Scarville voted 30-0 to renew the Cooperative's franchise.

—

M.J. Madson and Alan Swenson are contacting land owners to obtain easements for laying cable 35 to 40 feet inside owners' fence lines so as to avoid conflict with subsequent road programs. The Cooperative reimburses the owners for any damages to crops.

—

Operations engineer Kenneth Lein and outside plant superintendent Marlin Senne attended a subscriber loop testing school Nov. 28-29.

—

The December *Dial Tone* reports that the Cooperative will hire about twenty college students for 1968 summer work burying cable. "Please write us and we'll send you an application. We will work about sixty hours a week."

Two college students, summertime employees of the RTA, wind up wire on a motorized reel on the truck. The wire is from telephone poles, and now becomes junk upon the conversion to underground transmission.

1968

In 1968, the Cooperative entered into a joint venture with the Phillips Petroleum company to treat about 400 miles of buried telephone cable with Phillips compound JR551, which contained marlex plastic and R55 gopher repellent. The repellent did discourage the gophers from gnawing through the polyethylene sheath to the copper sheath, but the project was shelved by Phillips after the Environmental Protection Agency banned the compound.

The Cooperative's 1968 annual meeting was March 27 in the Forest City Civic Auditorium. Manager Bergland's financial report for 1967 listed revenues of $391,038.04 and expenses of $311,972.84, leaving $79,065.20 to be applied on the REA loan. The expenses included $80,407.77 reserved for depreciation and $27,456.43 loan interest.

Unpaid REA loans totalled $1,520,657.63.

The Cooperative had cash on hand on December 31, 1967, of $11,484.55, but reserves invested in bonds of $379,367.89. These figures indicate a keen avoidance of idle dollars.

After the meeting, the Cooperative's board of directors re-elected R. E. Aukes, President; Gustaf Gustafson, Vice-President, and Arlo Carson, Sec.-Treas.

Chile Assignment

It was announced at the meeting that Manager Bergland had accepted a one-month assignment to the South American country of Chile. There he worked with the rural electric cooperatives in the fields of sound financial planning and the creation of a financially strong cooperative federation. He also worked with the Chilean government to finance the lines of the rural electrification systems in Chile, as well as in exploring the possibilities of rural telephone systems.

Manager Bergland was chosen for the assignment by the U. S. Agency of International Development of the U. S. State Department. Chilean trip expenses were paid by the U. S. government.

In June, 1968, the Cooperative began construction of new switchboard buildings in Hanlontown, Joice, Leland , Scarville and Rake. The buildings, designed by Manager Bergland, were constructed of cement block, with electric heat in the concrete floors. They were—and are—equipped with air conditioning to dissipate the heat generated by the equipment during the summer months. They have air pressurization controls to filter the outside air and keep the buildings dust free. The buildings are completely fireproof and qualify for reduced insurance premiums.

In October, 1968, the old Emmons switchboard was installed in the Hanlontown exchange to provide the subscribers with dial service. New dial phones replaced the old magneto phones.

From latter November, 1968, to December 20, Manager Bergland took a second one-month assignment in Chile to assist in the Chilean government's rural electrification project in southern Chile. The assignment was under the auspices of the United States Agency for International Development and the National Rural Electric Cooperative Association. Besides helping the government there establish better financial and management practices, Manager Bergland also implemented procedures for the acceleration of Chile's electrification program.

Briefs - 1968

There is a saying around the Co-op that "when we put the plow in the ground, we feel we are halfway through with the job." This is because of the enormous amount of planning and paper work that must precede actual construction. Documents and papers of all units must list a complete inventory for all types of plant to be installed. All construction work is completely engineered on paper; every mile must show the exact number of cable pairs, terminations, pedestals, etc., so material can be ordered. Cable routes must be inspected prior to staking, after which detailed sheets are prepared on every half mile of construction. Permits must be obtained from town, county and state governmental bodies, as well as from highway and railroad jurisdictions.

—

Andrew Hove and Wayne Nyguard became employees of the Cooperative in the spring.

—

The Cooperative is selling telephone poles salvaged from the conversion to buried cable, priced as follows:

Length	Price	Length	Price
12′	$3.50	20′	$ 7.00
16′	5.00	25′	8.00
18′	6.00	30′	10.00

The poles are southern yellow pine, which is stronger than northern pine. They are treated with eight pounds of No. 1 creosote.

—

On August 14, Manager Bergland and state senator John Buren and state representative James Klein appeared at a hearing in Des Moines to protest the Iowa utility property tax law. The group pointed out that in Minnesota the Cooperative pays four percent of gross receipts in lieu of a property tax. The Co-op's Iowa property is assessed at twenty-seven percent of original cost, with no depreciation allowance. This resulted in the Cooperative paying a tax of $33,000 for 1968—more than ten percent of its gross revenues. The Cooperative's complaint emphasized that its revenue is low per dollar of investment because its service is provided to relatively thinly populated areas, unlike the dense customer base in cities. A gross receipts tax like Minnesota's is fairer because in effect it adjusts the value of a property to its earning capacity, the complaint pointed out.

An RTA derrick truck with power winch has just removed a telephone pole in the RTA's conversion to an all buried system. The folks in the house now can look at their trees sans telephone poles.

1969

In January, 1969, Clarence Willand was appointed to the board of directors of the Winnebago RTA to fill the vacancy created by the resignation of Norman Larson.

Mr. Willand lives three and one-half miles north and east of Joice.

The Cooperative's financial report for 1968, presented at the Cooperative's annual meeting March 26, 1969, listed $2,037,765 as owed by the Cooperative to the REA for loans and interest since 1951, net of loan payments since then of $657,181.93.

Revenues for 1968 were $423,281.44 and expenses were $351,528.33, including REA loan debt interest of $26,406.41, leaving $71,753.11 to be applied on the REA loan.

Following the annual meeting, the board re-elected R. E. Aukes, President; Gustaf Gustafson, Vice-President, and Arlo B. Carson, Sec.-Treas.

On March 20, 1969, the REA approved the Cooperative's loan application for $865,000, bringing to $3,900,000 the Cooperative's total borrowing from the REA for rebuilding and modernizing 17 exchange areas.

In April, 1969, the Cooperative purchased land on the east edge of Lake Mills from Harry Helgeson as a site for a garage and maintenance facility for the eastern half of its territory.

The conversion to private line service for Thompson was estimated at $280,000 and for Lake Mills rural at $120,000. The Cooperative had substantial savings by using its own crews as opposed to contracting.

In November, the Cooperative's crews began the installation of the new Thompson dial switching center, prior to conversion to private line service scheduled for March, 1970. It took three men approximately four months to complete the installation, wiring and testing.

Contract Frustration with General Telephone

As of November, 1969, the Cooperative had been unable to secure a satisfactory contract with the General Telephone Company for private lines switching service through the Forest City and Buffalo Center switchboards for the Forest City rural and Buffalo Center rural subscribers. Consequently, conversion to private service was being delayed.

The following article appeared in the November *Dial Tone*:

"During the past three years we have been negotiating with the General Telephone Co. for a new switching contract to give our subscribers one party telephone service in the Buffalo Center and Forest City rural areas. Most of you realize that the dial switching equipment in Buffalo Center and Forest City is owned by General Telephone Company. We have a contract with them to handle switching on our lines serving the rural subscribers in those areas. It had been our hope that a private line switching fee could have been negotiated at a price that would not have been excessive.

"The RTA-Lake Mills contract, which was executed over two years ago, made it feasible for the RTA to provide private line telephone service to the 187 RTA rural Lake Mills subscribers.

"The Lake Mills contract is twenty-one dollars a year or $1.75 a month per subscriber less than that asked by the General Telephone Co., or a total difference of $3,927 a year in favor of the RTA in the Lake Mills exchange.

"There are 238 RTA rural subscribers in the Buffalo Center area and 368 RTA subscribers in the Forest City rural area. The General Telephone Co. is demanding a $21 per year subscriber higher switching fee in Buffalo Center and Forest City than that which RTA pays the Lake Mills Telephone Co. This excess charge would amount to $4,998 a year more for the Buffalo Center rural patrons and $7,728 a year for the rural Forest City subscribers, of a total of $12,726 a year more than the Lake Mills rate."

Briefs - 1969

Rhonda Ramsey, ninth grade student who is confined to her home four miles south and west of Joice with multiple sclerosis, is keeping up with her Lake Mills high school class work by using a two-way radio telephone installed in her home by the Cooperative.

—

REA/RTA employees Darwin Adams, Adolph Swenson and Alan Swenson were awarded graduation certificates this January after completing a basic electricity correspondence course sponsored by the Winnebago REA/RTA under the auspices of the Rural Electrification Administration.

—

The June *Dial Tone* advertised real estate in Leland for sale by the Cooperative—the south five feet of lot 9, and lots 10, 11, 12, block 3, south of the switchboard building. Sale subject to release of mortgage by the U. S. government.

—

RTA employees Marlin Senne, Foster Hartman and Norman Risius have attended a two-day training school for the maintenance of automatic electronic direct distance dialing equipment.

—

The Joice exchange was cut over to private lines on January 5 and Leland on February 23.

—

At midnight April 3 the Scarville telephone exchange was cut over to all private line service, and at midnight May 10, Rake was cut over, Rake's mayor, Jerome Halvorson, making the first direct distance dialed call, to his sister in the state of Washington.

Darwin Adams operates a small tractor with a power take-off that rotates a drill pipe to bore a hole for telephone cable underneath a hard surface road, out of sight at bottom of picture. He is aided by a student summertime worker. When the Winnebago telephone and electric cooperatives terminated their joint operating contract at the end of 1973, Darwin remained a full-time employee of the Winnebago REC.

1970

At the Cooperative's annual meeting March 25, 1970, Manager Bergland reported 1969 revenues of $491,659.55 and expenses of $414,871.12, leaving $76,788.23 for REA loan payments, dividends and surplus. As of December 31, 1969, the Cooperative's indebtedness to the REA stood at $2,451,038.27.

The Cooperative's fixed assets—poles, lines, telephones, buildings, central office equipment, tools, construction equipment and construction in progress—totaled $2,235,015.09.

Manager Bergland was re-elected a director of the Iowa Telephone Association at its annual meeting, May 8, 1970, and was continually re-elected until his retirement in 1982. The ITA's 170-member telephone companies then served more than 800 Iowa exchanges with more than 1,400,000 telephones. Attending the meeting with Manager Bergland were Kenneth Lein and Marlin Senne.

In June, 1970, the Cooperative awarded a contract to Henningsen Steel Products Company of Mason City for a $34,000, 50'x100' steel garage and maintenance building in Lake Mills to house office, equipment workroom and storage for supplies and heavy equipment.

The building was completed in 1971. It was fully insulated and featured electrically heated concrete floors. It had a teletypewriter hookup with the RTA Thompson office to enable the crew members in charge of the eastern half of the Cooperative to receive and transmit information to the office in Thompson.

Manager Bergland was a founding member of the board of directors of the Consolidated Communications Corporation (CCC) organized in the fall of 1969. CCC was formed to build and operate a microwave system to furnish dependable low-cost audio, video and data transmission facilities to the State, estimated to cost $15 million.

The CCC founders were principally independent telephone companies and their owners. The new corporation was brought into being after the state legislature had appropriated funds for a coordinated and comprehensive study of the state's communications needs. The study recommended a network of approximately 45 microwave towers and associated equipment to handle educational TV, and other communications needs, including data transmission between centralized data processing centers and various state, county and local offices.

Heavy users, in addition to the ETV network, would be the Board of Regents institutions, civil defense agencies, highway patrol, and crime prevention agencies. Facilities would serve both stationary and mobile units.

The CCC was superseded by the state's building its own microwave system.

Briefs - 1970

On March 14, the Thompson exchange was cut over to private line service. Mayor Leonard Helgren made the first call, to his daughter in Wichita, KS.

—

The Cooperative is having problems with buried cable in the Grafton and Hanlontown areas due to frost damage because the cables were buried too shallow back in the early sixties by outside contractors prior to the time when the Cooperative switched to using its own crews to achieve better quality and more economical construction.

—

The Cooperative is burying cable in conduits under I-35 highway at a cost per crossing of about $2,000, which is refunded the Cooperative by the Highway Commission.

—

In June, the RTA outside employees were recipients of a national safety award, presented by Manager Bergland to outside plant superintendent Marlin Senne, for completing a full year's work schedule without a lost time accident.

—

The Cooperative is installing emergency power generators in all its switchboard offices. The new automatic generators will have a time delay on starting in the event of a power failure and outages; also, these generators will exercise themselves once or twice a week, as programmed. Installed in each generator is a crank case oil heater to insure starting during extremely cold weather.

—

The cost of communicating between two cities such as Chicago and Los Angeles by first class mail has risen 300 percent since 1932, and the possibility exists rates may by raised again to $.08 AN OUNCE. Telephone costs, in contrast, have dropped dramatically over the same period, and today the cost of a three minute daytime call between these cities is about one-fourth the cost in 1932. The postal service was one of the first and most important functions of the federal government when the Republic was founded. Telephone service, on the other hand, has been left in the hands of private enterprise. One can only wonder what would have happened if our founding fathers had telephones and had decided that only the government could make them work.

Ronald Reagan, Governor of California, as quoted in the May *Dial Tone*: "Some 25 years ago you could make a long distance call from San Francisco to New York for twenty-eight dollars. For the same amount of money you could send 1,368 letters. Today you can make the same telephone call for two dollars and a half, and for that amount you can only send forty-one letters. So the government is investigating the Bell system."

This 1969 photo shows Ken Lein doing the preliminary staking for a buried cable route.

1971

At the Cooperative's annual meeting in March, 1971, Milford Helgeson was elected to the Winnebago RTA board of directors. Mr. Helgeson farms north of Lake Mills.

At the meeting, the RTA members voted favorably on two amendments to the Cooperative's By-laws and Articles of Incorporation: (1) to increase the term of the board members to three years instead of two years, and (2) to limit board members to three consecutive terms. Number (2) was rescinded by the members a few years later.

At the meeting, the movie, "The Telephone and the Farmer," was shown. The movie was a 1951 production of the national REA, filmed in Thompson and around Lake Mills, with an all-local cast. The movie was then nationally distributed to show other areas how, by following the Winnebago RTA's methods, they could consolidate small telephone companies to make a larger operating unit for qualifying for an REA loan. Those living of the movie's cast were invited to be guests of the showing.

Manager Bergland's financial report for 1970 showed revenues of $579,617.62 and expenses of $482,753.55, leaving $96,960.35 for loan payments, dividends and surplus. The expenses included $48,171.76 REA loan interest and $113,743.56 reserved for depreciation.

In May, 1971, Manager Bergland was elected first vice president of the Iowa Telephone Association.

National Rural Telephone Bank

On September 15, 1971, Manager Bergland was appointed by President Richard Nixon to the thirteen member Charter Board of Directors of the newly created National Rural Telephone Bank in Washington, D. C.

The appointment was confirmed by the U. S. Senate.

The purpose of the Rural Telephone Bank (RTB) was to supplement the Rural Electrification Telephone Loans program, whose loan applications were greatly in excess of the money allocated to the program by Congress.

The RTB was controlled by its 13-member board, of which the REA Administrator served as governor. Four board members were officials of the U. S. Dept. of Agriculture, two were from the general public, and six, including Manager Bergland, were from operational telephone systems throughout the nation.

The presidential appointments were for one year only. In subsequent years membership on the board was by election by the nation's telephone companies.

Originally, the Bank was funded by purchases of its stock by telephone systems and borrowers, and by the government, which purchased $30 million of stock annually. These funds were then blended with capital borrowed by the Bank from the nation's large commercial banks. Such blending of funds enabled the Bank to give reduced interest rates based upon the borrower's financial needs.

Later, however, Congressional legislation made it possible for the Bank to borrow its funds directly from the U. S. Treasury at an interest rate slightly above the cost of money to the government. This fluctuating rate, adjusted in part by the borrower's ability to pay, determined the rate of loan interest, which ranged from four percent to ten percent and more.

The National Telephone Cooperative Association, of which Manager Bergland served two terms as president, had, during his tenure, advocated the creation of a bank to supplement the loan requirements of telephone systems over and above the funds available from the REA loans program.

On September 14, 1972, the first election was held for membership on the thirteen member board of the National Rural Telephone Bank, the thirteen incumbent members having been appointed by President Nixon upon the creation of the Bank in 1971.

Manager Bergland and Harold Payne of Export, Pennsylvania, were the only two of the original appointed members to survive the election and be retained on the board.

Of the 820 eligible rural telephone companies in the Bank, 732 voted in the election and the competition and in-fighting for board positions were fierce. Four ballotings were required to determine the election, although Manager Bergland was elected on the first ballot. Fifty-one percent of those voting, or 367 votes, was required for election and Manager Bergland received 475 votes on the first ballot.

Manager Bergland and Mr. Payne were re-elected to the board each of the nine succeeding years, until 1982, when they retired from the telephone industry.

During his last several years on the Bank's board, Manager Bergland served as the official Bank secretary, responsible for all the minutes of the board meetings.

In less than a year, by February, 1973, the Bank had loaned $177 million to 153 rural telephone systems in the U. S. A.

(Interesting footnote: In October, 1971, Manager Bergland reported to the RTA membership that because of the Cooperative's financial success achieved over the years, the Cooperative no longer was eligible for the two-percent-interest REA loans, and that it was therefore imperative that the Rural Telephone Bank become a viable source of loan funds for the Co-op.)

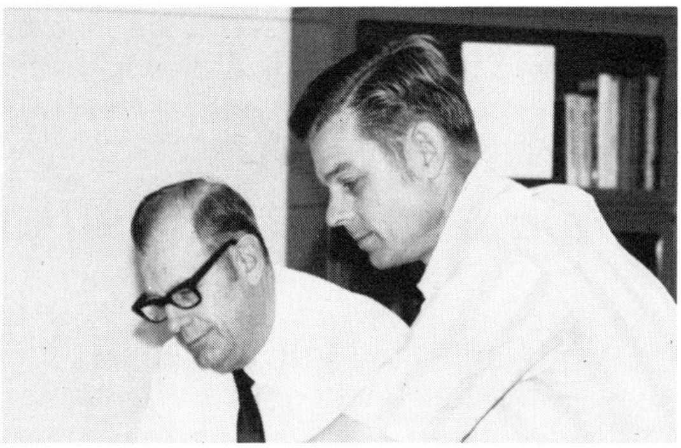

This 1971 photo shows Manager Bergland with Ken Lein of the Co-op's engineering department: Compatible associates!

Briefs - 1971

On June 7, Daryl Stelter accepted employment with the Cooperative, after completing a two-year electronics training course at North Iowa Area Community College.

—

Subscribers were glad the Co-op had installed automatic direct distance dialing, which does not require an operator, when the Bell Telephone System operators and craftsmen went on strike in June.

—

Humor in the July *Dial Tone*:

"A Thompson businessman had a thorough physical examination. Afterward, the doctor told him he was in good shape, but that his blood pressure and cholesterol count were a little high. This opinion cost money. He then called a cab to get back to work and got one of those friendly, talkative drivers.

'Been to the doctor?' asked the cabbie. The man said that he'd been in for a checkup. 'Well,' said the cabbie reassuringly, 'you look like a healthy businessman to me, except you probably have high blood pressure and your cholesterol count may be a bit high."

—

The Cooperative is building a 40' x 60' electrically heated maintenance facility in Thompson. Henningsen Steel Co., Mason City is the contractor.

—

The Cooperative employed eighteen college students in field work this summer.

—

A 43′ x 104′ unheated storage building was built this fall in Lake Mills, adjacent on the east to the Cooperative's maintenance building.

—

Manager Bergland appeared before the Buffalo Center City Council November 14 and requested their consideration of an election to grant the Cooperative a franchise to construct its own building and switchboard within the city limits as an alternative to the ongoing negotiations with General Telephone for the use of their switchboard for the Co-op's rural Buffalo Center subscribers.

—

In a speech to Worth County Farm Bureau members at their annual meeting October 6, Manager Bergland warned against the apathy of citizens who do not exercise their right to vote, citing as an example of such laxness the communist take-over of Chile, from which country Manager Bergland returned two years ago from a government assignment.

—

A sleet storm March 18 brought down many of the Northwestern Bell Telephone toll lines south of Forest City and as a result long distance service was disrupted in the exchanges of Buffalo Center, Rake, Scarville, Leland, Lake Mills, Joice, Fertile and Hanlontown. Also, 100 electric lines of the Winnebago REA were burned off at different places due to arcing between conductors caused when the ice melted off the bottom wire first, due to ground temperature and position of neutral wire next to the ground, which then came in contact with the top wire, causing a welding arc. By noon, Manager Bergland had gotten assurance from several electric crews throughout the state that they would come to the rescue; and at 1:00 a.m. that night, ninety percent of the electric system was restored to service.

—

The Cooperative awarded $25 savings bonds to each of three Co-op members who submitted winning designs for the cover of the Cooperative's annual report. Winners were: C. E. Garst, Buffalo Center; Barbara Purcell, Fertile; Vicki Trostheim, Scarville.

—

The Cooperative has hard-of-hearing telephone sets for noisy areas or people with hearing problems, for an additional charge of $1.25 per month.

1972

On March 23, 1972, the Cooperative appealed to the Iowa Commerce Commission to force Northwestern Bell Telephone to grant the Cooperative toll connections and the General Telephone Company to grant interconnection lines with switchboards which the Cooperative planned to install in Buffalo Center and Forest City.

The reader will recall that in 1950-51, Manager Bergland and Attorney Weible were able to hammer out a switching contract with General Telephone Company whereby General's switchboards in Forest City and Buffalo Center would be used for switching the Cooperative's Buffalo Center rural and Forest City rural subscribers.

In 1972, the Cooperative decided to install its own switchboards in Forest City and Buffalo Center for its rural subscribers in the two towns. The decision was made because negotiations with General Telephone for a contract that would give private line service to the two groups of rural subscribers had reached a stalemate. The Cooperative had such a contract with the Lake Mills Telephone Company for its Lake Mills rural subscribers for a rate of ninety cents a month, whereas General was demanding $2.65 per month; $21.00 a year more for each subscriber.

On March 23, 1972, the Cooperative appealed to the Iowa Commerce Commission to force Northwestern Bell Telephone to grant the Cooperative toll connections, and General Telephone to grant interconnection lines with the Cooperative's proposed switchboards.

The Commission ordered a hearing to be held October 17, 1972, on the Cooperative's appeal, but before that date the Commission canceled the hearing and ordered that the matter be further negotiated by the two parties until a settlement was reached.

(The Cooperative's appeal and the Commission's response were published in the *Dial Tone* of September, 1972. See facsimiles on pages 118 and 119.)

Here is Cooperative's Appeal . . .

Winnebago RTA serves 614 telephone subscribers in the Buffalo Center and Forest City rural areas. The Winnebago RTA, after negotiations with General Telephone Company, has decided to construct its own switchboards to serve the Buffalo Center and Forest City rural areas.

On March 23 the RTA appealed to the Commission to force Bell to grant us toll connections and the General Telephone Company to interconnect with our facilities. The RTA felt that this action became necessary due to a stalemate in negotiations over the past 4-5 years. The following is a copy of our resolution filing an appeal with the Iowa Commerce Commission and also an excerpt of the Commission's Order on Aug. 3 setting forth a hearing date on October 17. We believe the two following articles are self-explanatory in our attempts to improve the grade of service of those two areas. (582-3631)

RTA Board Resolution to Commission

To Iowa Commerce Commission

March 23, 1972

In behalf of the Board of Directors and Manager Glenn Bergland, I am submitting the following resolution adopted at our Board of Directors' meeting on March 23, appealing to the Commission for your assistance in a solution of the following problem.

It was moved and seconded that the following resolution be adopted:

Whereas, the Winnebago Cooperative Telephone Association has been proud of its service to its telephone subscribers, and six years ago embarked on a program of trying to convert all its exchange areas to all private line systems and

Whereas, our Association has already converted ten of our seventeen exchange areas to an all private line system, two more will be converted this spring and another to be converted next fall, and

Whereas, this will mean we will have only two exchange areas with all aerial plant with ten or less subscribers on each line, viz., the rural areas of the Buffalo Center and Forest City Exchanges and

Whereas, prior to the creation of the Winnebago RTA Association, there was a farmer switchboard in Forest City and a town switchboard with an interconnection between the two and

Whereas, at the time of our consolidation in 1950, we proceeded to negotiate a switching contract to modernize our system with the Central Telephone Co., the predecessor of the General Telephone Company, serving these two towns and

Whereas, by said arrangement we converted these areas to a dial system and

Whereas, our Association now wishes to convert these rural exchange areas to an all private line system and

Whereas, during a period of four to five years, we have attempted to negotiate with the town company to obtain a satisfactory switching contract enabling us to obtain financial feasibility to make said conversions and

Whereas, all said negotiations have proven fruitless and

Whereas, the Association has explored alternate proposals by which we can resolve our conversion on an economic feasibility basis and

Whereas, our studies show that we can effect approximately a $30,000 differential in income by the establishment of a second switchboard in each of these areas and still not degrade from the service of any subscriber and

Whereas, this is the difference between a deficit and a profitable operation and

Whereas, in order to effect the establishment of the second switchboard, it becomes necessary for us to obtain a toll connection from Northwestern Bell Telephone Company, either at our exchange or that we meet the Bell Co. with our cables at another point, and in addition to this toll connection we will want to have connection agreements with General Telephone Company so as not to degrade the service through the interconnecting of trunks between the two exchanges and tandeming EAS facilities and

Whereas, the Bell Company has, by their letter of March 20, 1972, a copy of which is enclosed, indicated that their legal department has serious doubts whether or not these toll connections can be made. Also enclosed is a copy of our response to Bell on March 23, 1972, indicating our desire to pursue this matter by every means possible to obtain this connection.

Now, therefore, be it resolved that the Secretary and Manager Glenn Bergland be hereby authorized to submit copies of this resolution to the Iowa Commerce Commission and

Be it further resolved that our Association makes an appeal to the Iowa Commerce Commission for their intervention in our behalf.

The Chairman called roll. All members present voted "Aye." The resolution was unanimously adopted.

Signed: Arlo Carson, Secretary
(896-2272)

. . . and here is Commission's Answer

Commission Orders Hearing

Iowa Commerce Commission
Order Setting Complaint for Hearing and Defining Issue Aug. 3, 1972

It has come to the attention of the Commission that a dispute exists among three utilities; Winnebago Cooperative Telephone Association (Winnebago), Northwestern Bell Telephone Company (Bell), and General Telephone Company (General).

Winnebago provides telephone service to the rural areas only of Buffalo Center and Forest City as well as eleven other Iowa exchanges. General provides the exchange service within Buffalo Center and Forest City. Winnebago and General have a switching contract for General's switchboard service from the town areas. Winnebago is considering the establishment of individual central offices at Forest City and Buffalo Center, so that it may provide private line service to its rural subscribers, since Winnebago and General are not able to agree on what would be a reasonable switching contract for General providing the necessary central office failicites.

In additon, there also appears to be a "three-way" disagreement between General, Bell and Winnebago on the wisdom of constructing a separate rural central office by Winnebago. This would require duplication of existing central office facilities as well as additional investment on the part of Bell and General plus a loss of revenue to General. (745-2947)

The Commission Finds:

1. Pursuant to Chapter 490A, Code of Iowa, 1971, we have jurisdiction over the service provided by these utilities.

2. Pursuant to Section 490A. 3, the Commission may, on its own motion, initiate a formal proceeding if a complaint has not been satisfied and if a reasonable ground appears for investigating a complaint.

3. Bell, General and Winnebago have not been able to reach agreement on the matter of rural central offices, and there appears to be reasonable grounds and good cause for the Commission to institute formal proceedings.

4. The issue on hearing is to be whether Winnebago should be allowed to construct rural central offices for its affected rural customers in Buffalo Center, Iowa and Forest City, Iowa and determine the best interests of telephone subscribers of the three companies. (588-3433)

The Commission Orders:

1. Hearing be held in this matter at the Commission offices in Des Moines, Iowa on October 17, 1972, at 10:00 A. M., in Hearing Room A on the Seventh Floor of the Valley Bank Building.

2. The issue considered at such hearing shall be in accord with the above findings in this matter.

Signed by: Iowa Commerce Commission Aug. 3, 1972

With further negotiations, the Cooperative and General agreed upon a multi-party switching rate of ninety cents a month per subscriber with a minimum of six subscribers on each line (total $5.40), and General granted the Cooperative a five-year option to acquire Buffalo Center town exchange in return for which General would acquire the Cooperative's Forest City rural system. The effective date of the option was January 1, 1973.

* * *

In April, 1972, Manager Bergland was re-elected first vice-president of the Iowa Telephone Association at its annual convention in Des Moines.

The Cooperative's annual meeting was held June 3, 1972, but the attendance was smaller than average because of the hot weather and field work.

Featured speaker at the meeting was David A. Hamil, Administrator of the federal REA and Governor of the Rural Telephone Bank.

Manager Bergland gave the Cooperative's financial report for 1971. Revenue was $590,758.11 and expenses were $507,319.37, leaving $83,438.74 for REA loan payments, dividends and surplus. Fixed and current assets totalled $3,235,951.87. REA loan debt was $2,300,332.27.

Mrs. Diane Rygh Nall of Corwith, granddaughter of the late M. J. Rygh, charter RTA board member, sang several songs at the meeting.

Following the meeting the Cooperative's board of directors re-elected R. E. Aukes, president; Gustaf Gustafson, vice president; Arlo B. Carson, Sec.-Treas.

On August 9, 1972, Kenneth Lein, Winnebago RTA's Operation's Engineer, was elected president of the Rural Iowa Independent Telephone Association at its annual meeting in Davenport.

Briefs - 1972

On January 21, the newly established Rural Telephone Bank, of which Manager Bergland is a board member (one of 13), made its first five loans, totalling $4,774,900, at interest rates ranging from 4 1/2% to 8% to three cooperatives and two commercial phone companies in Texas, Kansas, Iowa (Sioux Center), North Carolina and Georgia.

—

"Prestafones," the numbers of which are pressed instead of dialed, are available to the Co-op's members for a five dollar installation charge and three dollars per month.

—

The Cooperative has scheduled open houses, including free pancake suppers, in its 17 exchanges during April, 5:00 p.m. to 9:00 p.m., to tour the dial equipment buildings.

—

Printed more than 200 years ago in "The Decline and Fall of the Roman Empire," by Edward Gibbon, and reprinted in the February *Dial Tone*, the following lines are as meaningful today—or more so—than they were in 1778:

"There are five basic reasons why that great civilization withered, was conquered, and died:

*The undermining of the dignity and sanctity of the home, which is the basis for human society.

*Higher and higher taxes, the spending of public money for free bread and circuses for the populace.

*The mad craze for pleasure, entertainment becoming every year more exciting, more brutal, more immoral.

*The building of great armaments when the real enemy was within—the decay of individual responsibility.

*The decay of religion; faith fading into mere form, losing touch with life, losing power to guide the people.

—

On April 7, the Crystal Lake exchange was cut over to an all private line system.

—

In June, the county road grader cut a 100-wire cable east of Emmons. Thirty-two and one-half man hours were required to make repairs.

—

The Cooperative's field workers are now wearing hard hats, per OSHA—Occupational Safety and Health Act.

—

On November 15, the Iowa State Office of Planning gave first place awards to three Winnebago RTA communities. In the Iowa Community Betterment category, Grafton won first place for communities under 500 for park and playground development, five units of low income housing, sewer system planning, and youth and beautification committee efforts.

Woden won second place for it library, low income housing, sewer survey and new subdivision for housing.

In the Community Development Project category, Thompson won first place in the 500 to 1,500 population bracket for its modern, heated swimming pool.

1973

The year 1973 was a reorganization year in the history of the Winnebago RTA.

Long-time president R. E. Aukes retired; Manager Bergland announced his retirement effective December 31, 1973, after which date the electric and telephone cooperatives would no longer have the same manager; the REA/RTA joint management contract was dissolved; the REA/RTA no longer would have the same members serving both boards; and the RTA headquarters was moved to Lake Mills from Thompson.

Mr. Aukes retired January 4, 1973. He was known as "Mr. REA" because of his leading role in the organization of the Winnebago Rural Electric Cooperative Association in 1938, and his active membership on the REA board (and RTA board). He served as president of both boards from 1953 until his resignation.

R. E. Aukes

As one of the original eight charter organizers of the Winnebago Rural Electric Cooperative in 1938, Mr. Aukes actively solicited his neighbors to get them to sign up for electric service from the cooperative. In those days electricity had to be sold because many farmers were not convinced they could afford the $3.75 per month minimum billing.

In 1939, the new cooperative obtained a federal REA loan for $258,000; the Winnebago REC was in business.

Again, in 1949, Mr. Aukes was a charter organizer, this time of the Winnebago Rural Telephone Association.

Mr. Aukes, 75 years old when he retired, remembers that when he was a boy of fifteen his father and a neighbor instituted the first phone system in the rural Buffalo Center area. The two farmers ordered telephones from the Sears catalog, connected them to the barb-wire fencing between the two farms, and set a time each evening when they would pick up their phones and talk to each other.

At the time of Mr. Aukes' retirement, thirteen of the Cooperative's telephone exchanges had buried cable and all private lines. Curiously enough, Mr. Aukes' own line in the Co-op's Buffalo Center rural system was still a party line, served by overhead wires.

Mr. Aukes pursues woodworking in his retirement and at age 92 flies his own plane.

* * *

On January 4, 1973, Gustaf Gustafson, REA/RTA vice president, was elected president of the REA/RTA boards of directors, replacing Mr. Aukes in that position.

Mr. Gustafson was a charter board member of the Winnebago RTA. He farmed northeast of Forest City, in the Mt. Valley township.

In February, 1973, Lloyd Kronemann, Grafton, was appointed to the RTA board by its members to fill the vacancy created by the resignation of Mr. Aukes.

Gustaf Gustafson, founding board member of the Cooperative, served 1950–1974; secretary-treasurer, 1957–1966; vice president, 1966–1973; president, 1973–1974.

Dissolution of Winnebago REA/RTA Joint Operating Contract

In 1971, Manager Bergland notified the Winnebago REA/RTA boards of directors of his intention to retire, giving December 31, 1972, as a tentative retirement date.

During the many board meetings that followed, Manager Bergland and the directors discussed how the cooperatives' administration would be constituted after his retirement.

It was decided that the joint management contract between the two cooperatives should be dissolved and that the cooperatives should have separate administrations and managements, and that because the telephone cooperative had expanded so far east, its headquarters should be moved to Lake Mills.

The increasing size of the telephone cooperative and the advances in technology making the electric and the telephone systems increasingly dissimilar became the reasons for their separation.

Because of the complexities of the mechanics of separation, the boards of directors asked Manager Bergland to delay his retirement for a year, until December 31, 1973, to give Manager Bergland more time to effect a smooth separation.

All assets and equipment of the two cooperatives, hitherto interrelated in their use, would have to be permanently divided between the two systems; the cooperatives would have to settle their accounts with each other; crew and office personnel that served both cooperatives would have to be assigned to one or the other; separate administrations would have to be established; a smooth transfer of RTA headquarters to Lake Mills would have to be brought about; and the joint management contract between the two cooperatives would have to be dissolved, with two new managers appointed to succeed Manager Bergland in the management of the two cooperatives.

A review of the joint management contract of the two cooperatives may be in order at this point.

Immediately following the January 15, 1950, RTA organizational meeting, a discussion ensued as to how the new cooperative would function and be administered.

Since the electric cooperative was a viable, going concern with resources and manpower, why not, the board and Manager Bergland decided, have the two co-ops enter into a management contract, whereby the Electric would manage the Telephone on an actual cost basis?

In line with this, it was agreed that the employees should not have two managers, nor should the manager work for two different groups. It was decided, therefore, that the same group of men should serve on both the electric and the telephone boards. Mr. Bergland would manage both of the cooperatives.

The board instructed Manager Bergland, assisted by attorney Dudley Weible, to draft a proposed management agreement. The agreement set out in detail the allocation of the manager's time, and the cost allocation of office overhead, utilities, crew payroll, tools and equipment, etc., etc.

There would be two separate boards of directors but the same men would serve on both boards.

The boards approved the management contract presented by Manager Bergland and attorney Weible. If there hadn't been such a contract, whereby the Electric backed up the Telephone, Telephone may not have survived its birth. During the first three years alone, the Telephone's expenses were so much more than its revenues that Telephone became indebted more than $100,000 to the Electric. The new Co-op's financial problems were exacerbated by the federal wage and hour law, which limited to forty hours the workweek of the Co-op's many telephone operators. Overtime pay was due after forty hours. Those same operators had been on duty day and night all week for the old pre-merger companies at a very modest salary.

Manager Bergland Sets Retirement Date

In March, 1973, Manager Bergland made a public announcement in the *Dial Tone* of his decision to retire, on December 31, 1973: "Almost thirty-five years have transpired since my first office consisted of a card table and a folding chair in a corner of County Agent Clark Mechem's office. The REA was only a paper entity. No electric lines had been built and we did not have a source of power.

"Beginning in 1950, we commenced the organization of the Winnebago RTA, which has resulted in the consolidation of sixty-seven telephone systems serving seventeen different exchange areas into one modern system. Our results have gained national recognition from the standpoint of operations as well as financial soundness.

"But there does come a time in everyone's life when it is necessary to turn the reins over to others. Personally, I believe it is better to make this decision when there is still time to readjust and to enjoy new endeavors with the blessings of health.

"It is a great feeling to be able to step down from an organization which has an excellent staff, a sound board of directors and a very cooperative and congenial membership.

"So to all of you members, I want to express a sincere 'thank you' for your wonderful cooperation and with the hopes that you will help us during the ensuing transition."

At the Cooperative's annual meeting on September 13, 1973, the members voted 1,989 to 619—a seventy-six percent favorable vote—to move the Cooperative's headquarters to Lake Mills from Thompson.

The Cooperative's board earlier had unanimously passed a resolution supporting the move, but membership approval was required for an amendment to the Co-op's Articles of Incorporation allowing the move.

Temporary partitions were installed in the Lake Mills garage building to house the re-located office personnel. The building was electrically heated and insulated and consequently was comfortable. The office crews and the office were moved over in December, 1973.

The decision to move followed earlier membership approval of the termination December 31, 1973, of the joint operating contract between the Winnebago Rural Electric Cooperative (Winnebago REA), in effect since the RTA was established in 1949.

On June 10, 1973, Manager Bergland was elected president of Iowa Telephone Association.

In the late summer of 1973, a second warehouse was built at the Cooperative's Lake Mills maintenance depot.

At the Cooperative's annual meeting in September, Manager Bergland reported 1972 revenues of $625,986.40 and expenses of $529,787.27, leaving $96,199.19 for loan payments, dividends and surplus.

At the meeting, Manager Bergland was presented with a plaque in recognition of his thirty-six years of service with the Winnebago REA and twenty-five years with the Winnebago RTA.

Manager Bergland, with Mrs. Bergland, receives a plaque of appreciation from R. E. Aukes at the 1973 annual meeting of the Cooperative. The occasion honored Manager Bergland's achievements as manager of both the Winnebago Rural Electric Cooperative and the Winnebago Cooperative Telephone Association since their inception, the Electric in 1938 and the Telephone in 1949.

The bronze plaque, which featured an engraved likeness of Mr. Bergland, was presented by R. E. Aukes, former president of the Winnebago REA/RTA.

Kenneth Lein New RTA Manager

In December, 1973, Kenneth Lein was appointed manager of the Winnebago RTA, to succeed Mr. Bergland on January 1, 1974.

"Kenneth Lein was a very ambitious and dedicated individual," said Mr. Bergland. "He was well qualified and worthy of his promotion to become manager."

Mr. Lein graduated from high school in Kanawha and began his employment with the Winnebago RTA in 1951.

He worked first with the outside crews, and later on with training and attendance at dial schools he became a switchboard technician.

Mr. Lein was then promoted to inventory control and work order clerk. He also began staking the lines for new construction. In addition, he helped engineer additions of the switchboards as they were needed.

Later, Mr. Lein assumed the responsibilities of staff assistant and operations engineer.

Briefs - 1973

On February 21, the Rural Telephone Bank's operation review committee, of which Manager Bergland is chairman, met in Washington to hear testimony from four large national associations as to their recommendations for policy and proposed legislative changes affecting the Bank.

—

The February *Dial Tone* pictured Daryl Stelter installing additional dial office equipment in the Fertile dial office building.

—

On March 23, the Fertile exchange was cut over to all private lines. The RTA's Foster Hartman made some 200 relay changes in the Fertile switchboard.

—

In 1972, the Cooperative received 736 reports of telephone "trouble," which turned out to be no-answer calls because nobody was home.

—

The RTA's Norman Risius and Foster Hartman are currently instructing a series of night classes for other employees in the maintenance of direct dial automatic number identifiers, and a second course in multi-line key telephone systems used by many RTA business customers.

—

In July, the Cooperative eliminated the twenty-five cents late payment penalty for telephone bills, but increased to $5.00 the fifty cents collection fee.

—

The RTA's Alan Swenson was pictured in the July *Dial Tone* tracing a case of trouble in an automatic subscriber number identifier.

—

In the annual issue of the *Dial Tone*, published in September, there is a picture of Wayne Nyguard seated at a desk pouring over "materials and records."

—

During the years 1939-1974, the Winnebago electric and telephone cooperatives helped sixteen of its employees to become systems managers of other associations. Generally, the cooperatives took the position that to attract good employees and keep them interested and motivated, there must be openings up the ladder to which they can be promoted. This required making room at the top by helping them find management positions elsewhere.

—

> "This is Perkins, Potter, Parker and Potts; good morning."
> "Is Mr. Potter there?"
> "May I ask who is calling?"
> "This is Mr. Sullivan's office, of Sullivan, Chadwick, Bicknell Jones."
> "Just a moment, I'll connect you."
> "Mr. Potter's office."
> "Mr. Potter, please. Mr. Sullivan wants him."
> "Will you put Mr. Sullivan on the line, please?"
> "Mr. Sullivan? Ready with Mr. Potter."
> "Hello, Pete? This is Joe. Okay for lunch. Good. See you."
> Alexander Graham Bell said it in fewer words: "Mr. Watson, come here, I want you."

—

Our annual meeting attendance record was set at the REA/RTA Cooperatives' annual meeting in March, 1973. So great was the crowd that after 1,959 of the Co-ops' silver dollars were given to members and spouses, the Co-ops had to revert to giving greenbacks.

—

In addition to the classes taught by personnel of the Cooperative to its employees (which include classes in electricity and telephony) the Cooperative instituted a correspondence course two years ago for employees, furnishing them with textbooks and correcting their test papers for them.

—

The cover of this year's Directory depicts the painting, "Fall Scene," courtesy of the artist, Floyd Thomas, Lake Mills, Iowa.

Glenn Bergland, Manager of the Winnebago Cooperative Telephone Association, 1949 through 1973, and Manager of the Winnebago Rural Electric Cooperative Association, 1939 through 1973.

1974

On January 1, 1974, Kenneth Lein began his duties as manager of the Winnebago RTA. He was appointed to the office by the RTA's board of directors upon recommendation of retiring Manager Bergland. Manager Lein was promoted to manager from the position of operations engineer and staff assistant.

The following letter from Manager Bergland to the RTA's subscribers was printed in the January *Dial Tone*:

"Dear Members: Congratulations to Mr. Lein on his appointment to become my successor. He has earned this opportunity and may we all wish him well in his new position.

"There is very little comparison between our RTA today and the sixty-seven 'whoop and holler' telephone companies with rings of 'two shorts and one long', etc. Back in 1949 and 1950, when we held numerous meetings to organize our RTA of today, we encountered many trying experiences and obstacles.

"The next step to dial and 'ten on the line with private ringing' was a great improvement. Then the development of plastic cables made 'all private line' systems feasible.

"Today we are proud of our RTA, its accomplishments, service, rates and our financial condition. We are one of the best in the nation, even though we mainly serve rural territory.

"Your cooperation, together with a group of fine dedicated employees, has helped make this possible.

"True enough, my wife and I are anxious to have some free time; but I will also miss your association and the responsibilities that you entrusted with me. But time passes on.

"May I wish you a rewarding new year.

Thanks to you,
Glenn Bergland"

Manager Lein and his wife, Janis, were then living in Thompson and were the parents of two sons and two daughters. Manager Lein was also president of the Rural Iowa Independent Telephone Association.

Kenneth Lein, Manager of the Winnebago Cooperative Telephone Association from January 1, 1974.

The passing of the leadership torch of the Cooperative from Manager Bergland to Manager Lein was another milestone in the history of the Cooperative. It is appropriate here to review the purpose and raison d'etre of a cooperative. Manager Lein states it well as follows:

"A cooperative is a business like any other business, differing only in the concept that it belongs to the people that use it. It is customer owned as opposed to investor owned.

"There are cooperatives of many types, including those in which the members use the services of the cooperative to market farm products, to obtain farm supplies, and to obtain credit for various enterprises; and of course, there are consumer service cooperatives, such as the electric and telephone cooperatives with which most of us are familiar.

"The purpose of a cooperative is to provide goods and services that cannot be provided at reasonable cost by someone else. Savings can be realized by a more meaningful production of these goods and services.

"It was nearly 230 years ago that the first cooperative was formed in the United States. Cooperatives help the communities they serve by adding to the financial base of the community, by providing jobs for local people who in turn buy goods and services and pay taxes that are used for schools, roads and myriad other community expenses.

"Any excess of the cooperative's revenues over expenses is returned to the members in the form of deferred dividends which may be paid in cash at a future date when finances permit.

"A cooperative is principally controlled by its user-members. This control, however, is to some extent limited by the law and rules and regulations to which a cooperative is subject.

"Many cooperatives grew out of the dissatisfaction of its members with prevailing prices and services. The telephone cooperative is a good example of this. The commercial phone companies avoided expansion into low density-low profit rural areas, choosing instead to serve high density, profitable urban areas. Hence, the joining together of individuals who formed their own subscriber-owned telephone company—the telephone cooperative."

* * *

It is also appropriate to set out the functions vis-a-vis the cooperative's board of directors and the cooperative's manager. Manager Lein illuminated these in a speech he gave in St. Louis to Regions 4 and 5, National Telephone Cooperative Association:

"A capable board of directors and manager are the backbone of a successful cooperative. In order to effectively carry out their responsibilities, each should have a thorough knowledge of the cooperative's policies and objectives. Neither should abdicate or give up undue authority to the other and by doing so burden the other with undue responsibility.

"Very simply, the job of the board is to establish basic operating policies and to employ effective management to carry them out. Since the directors as a body are legally responsible to the members for the cooperative's management, this is a serious and important responsibility.

"The cooperative's manager is responsible for carrying out policies established by the board, for directing the day-to-day operations and keeping the board advised at regular board meetings, and for working with the board to achieve its objectives.

"Crucial to this relationship is a clear understanding between the board and the manager regarding the authority and responsibilities of each.

"The flow of authority must pass through the manager. He is the connecting link between the board and employees. If cooperative directors begin giving orders or directions to cooperative employees, the manager's authority with the employees will quickly be diminished or disappear completely. One boss is enough.

The Cooperative's board of directors, 1990. Left to right: Jerry Epland, Keith Finer, Milford Helgeson, Vernon Jacobson, Lloyd Kroneman, DeVere Steffensen, James Petersburg.

"Further, as a supplement to high quality telephone service, a plan to help maintain the image of the cooperative should be conducted on an on-going basis, realizing that favorable customer relations and good will are cumulative within the cooperative organization and can be drawn on when loyalty and support are needed most. Each employee should be charged with a share of this responsibility and each member of the cooperative board has a responsibility to become an expert on cooperative business principles, policies, finances and programs so as to serve as an effective salesman for the cooperative. If he shirks this responsibility by depending too heavily on the manager and employees to promote the cooperative, the job will get only half done at best.

"Cooperative managers normally have the authority and responsibility to determine objectives, establish procedures and produce results within the scope of policies set out and delegated to him by the board. He should also act as an advisor to the board with respect to numerous areas of board activity and responsibilities.

"One example of this is the retirement of capital credits. The practice of making capital credit retirements offers a benefit to cooperative member-owners that other businesses do not provide. Without exception, telephone cooperatives that make regular retirements of capital credits have found such to be an extremely effective means of developing support. They provide a tangible benefit of cooperative service while increasing member education and understanding of cooperative non-profit operations.

"Every dollar used to retire capital credits is a dollar spent for solidifying and increasing member support.

"Cooperatives have always relied on the trust, loyalty and support of their membership, and this becomes even more important as we face the future in our challenging and dynamic industry.

"Again, if managers do not fulfill their obligation to advise directors, the cooperative's efforts may fall short of their objectives. Managers and directors should work together to develop a clear understanding of their authority and responsibilities. Doing so will result in a smooth, effective board-manager team."

* * *

One of Manager Lein's first responsibilities, as assigned by the board in January, 1974, was to complete plans and specifications for the construction of the Cooperative's headquarters building, which stands today on the east edge of Lake Mills. The building was completed in March, 1975, and office staff and furniture were moved into it from their temporary location—since December, 1973—in adjoining warehouse and garage. The Cooperative acted as its own contractor, with a resultant savings in construction costs. The Cooperative also did its own excavating for the building.

Manager Lein checks a blueprint in construction of the Cooperative's headquarters building, Lake Mills. The Co-op was its own contractor. (1975)

With the operational split of the RTA and REA, it was clear that the RTA should have its own radio system, and after extensive negotiations and bids, a base system and nine mobile radio units were purchased from Motorola for $24,000. Additions and modifications have been made and the system is in operation today.

In his January, 1974, report to the board of directors, Manager Lein said that $35,433 was saved through the use of the Cooperative's own men and equipment to bury 113 miles of cable in the conversion of Buffalo Center to private line service, completed in November, 1973. The Cooperative's actual construction costs, including labor, material, overhead, depreciation and equipment, was $132,217, as compared with

RTA's maintenance crew, 1973. Left to right: Norman Risius, Charles Hunt, Marlin Senne, Steve Kallestad, Jim Simon, Gary Yost, Wayne Roberts, Daryl Stelter. (Foster Hartman is missing from picture.)

contractor estimate of $167,650. Sixteen college students were employed during the summer work.

"An added benefit is assurance of the highest quality of work when we do it ourselves," Manager Lein said.

With more than eighty percent approval of private line service by the citizens of Grafton and Woden, the Cooperative contracted in January, 1974, for the purchase of the dial office equipment required.

The Cooperative, at that time, made the decision to use its own technicians to install the new dial switching equipment as opposed to paying Stromberg Carlson for the work as was done previously.

The cable for the Grafton/Woden private lines was buried twenty to thirty feet back of the fence lines, where it could not be damaged if roads were widened, ditches dug out for tiling, culverts, driveways, etc.

Reimbursements for Crop Damage

The procedure by which the Cooperative reimbursed land owners/operators for buried cable crop damage was as follows:

A Cooperative employee counted and measured the rows of crops damaged and then reached an agreement with the owner/operator on the yield involved and the price.

In the case of buried cable for Woden and Grafton, the Cooperative paid crop damage of $10,011—about $100 per mile.

"We are always especially pleased with the reception we get from our agricultural people, our farm operators and tenants on the installation of cable behind the fences," Manager Lein said.

Early in 1974, the Lake Mills Telephone Company began conversion of its crossbar switching to new electronic transistorized dial equipment. Because the new equipment is especially sensitive to lightning and other induction interference, the Cooperative's board made the decision to bury its Lake Mills rural aerial service which then was being switched through the Lake Mills exchange. In addition to removing the interference threat to the equipment of the Lake Mills exchange, the buried cable protected the rural system from the hazards of weather.

In April, 1974, Winnebago RTA board member DeVere Steffensen was named a "Master Farmer" by *Wallace's Farmer* magazine. Mr. Steffensen, who has served on the Winnebago Board of Directors since 1967, farms 240 acres northwest of Leland, Iowa.

The Cooperative's annual meeting was held September 19, 1974. Manager Lein reported to the membership that "working with our legislators and regulators, and the IRS, we were able to prevent the IRS from changing our cooperative and others to a federal taxpaying status from our current exempt status, the latter requiring that at least eighty-five percent of our revenues must come from our members."

Manager Lein also reported 1973 Cooperative revenue of $684,207.51 in expenses, including depreciation and retired plant amortization, of $579,414.17, leaving $104,793.34 for loan payments, dividends and surplus. As of December 31, 1973, the Cooperative's assets totaled $3,598,676.73, and its REA loan balance was $2,452,152.61.

"Your cooperative saved over $600,000 in the past seven years by doing its own construction with plows and equipment owned by the Cooperative and operated and supervised by our own employees," Manager Lein told the members.

Following the meeting, the board of directors re-elected Gustaf Gustafson, President, Arlo B. Carson, Vice President, and Victor D. Knutson, Secretary-Treasurer.

Briefs - 1974

Mr. and Mrs. Glenn Bergland were honored January 3, 1974, by the RTA and REA employees at a dinner at Clear Lake, Iowa, in recognition of retiring Manager Bergland's thirty-five years of service to the two cooperatives.

—

The Cooperative sold an old North-Electric automatic dial switchboard and the wood frame building housing it to the Miller Telephone Exchange for $1,500, making it possible for the Miller exchange to dispense with its operator board. The switchboard had served the Scarville exchange from the early Fifties to 1969. Miller moved the building as a condition of the sale.

—

RTA employee Alan Swenson conducted a survey in February in Grafton and found that more than eighty-five percent of the Grafton residents favored an all private line system.

—

Cable construction will be completed this Fall for single party service for Grafton and Woden with cutover scheduled for Spring, 1975. Alan Swenson is in the process of obtaining easements from landowners for burying cable.

—

Cooperative technicians Foster Hartman and Norman Risius will install the switching equipment in the Grafton and Woden exchanges.

—

The Grafton and Woden exchanges were exempted from the $1.50 a month increase in multi-party rates effective the first of July because the two exchanges are not yet converted to single-party service.

—

The September *Dial Tone* reports that the RTA's new Lake Mills office building will have walls of cement block with a four-inch cavity and be faced with brick on the outside. The cavity will be filled with urethane insulation, which has twice the insulating value of styrofoam.

—

The Cooperative's board of directors holds its meetings in the basement of the Farmers and Merchants State Bank in Lake Mills, pending the completion of the new office building.

—

The cover of this year's Directory depicts the paintings, "Pilot Knob Tower," and "Fertile Mill," courtesy of the artist, Viola Buckley, Fertile, Iowa. "Fertile Mill" was painted from a photograph of the mill taken in 1905.

—

Phone Deposits

Periodically, the Winnebago RTA writes off bad debts—unpaid telephone bills. In 1974, for example, the Cooperative's bad debts totaled $1,757.00.

"When a telephone bill is not paid and must be written off as uncollectible, it becomes an operating expense that must be borne by all the customers," Manager Lein explains. "To keep this expense down, we institute revenue protection measures, principally by taking deposits.

"We don't like deposits and we try to reduce the number we take, but there are some things to keep in mind," Manager Lein points out.

"The Cooperative is a public utility and by Iowa and Minnesota law it must provide service unless the customer owes the Cooperative money. This means that everyone in, or moving into, the Cooperative's exchange area has the legal right to a telephone unless that person owes the Cooperative for past service.

"But if that same customer walked into a department store and applied for a charge account, after running a credit check, the store could reject the application. In such case, the customer could still pay cash or apply for credit at another store. Such options, however, are not available to the telephone customer; nor is credit refusal to a new customer in the power of the Cooperative.

"The Cooperative, however, can and does make a credit check with other phone companies of which the customer has been a patron. After reviewing the customer's credit record—good, bad, or no record—with the other phone companies, the Cooperative will make a decision on whether or not to require a deposit.

"The customer needs to understand," Manager Lein continued, "that with every new telephone, the Cooperative extends at least forty-five days of unrestricted long distance credit. Thus, the Cooperative is in the credit business in a big way. For the most part, the Co-op has no way of screening account activity during the month to be sure it is not being misused. It would undermine our basic service of speed and convenience if we stopped before each long distance call and said, 'pardon us, sir or madam, but we have to check your credit rating.' "

Manager Lein points out that "if the customer has a good credit history, we don't ask for a deposit.

"If the credit history is not satisfactory, then we take a deposit which usually is twice the total of local service and estimated long distance calls that could occur on the customer's bill, with a minimum of fifty dollars."

This 1972 photo shows the RTA's Ken Lein pouring pancake batter at one of the open house pancake days the cooperative featured that summer at each of its exchanges.

1975

On December 31, 1974, Gustaf Gustafson resigned from the Winnebago RTA board of directors and as president of the board. Mr. Gustafson, Forest City, had served twenty-five years on the board and was one of the original organizers of the Cooperative in 1950. He was board vice president for twenty-three years and then president until his resignation.

The board of directors elected Arlo Carson as new president from his position of vice president, and elected board member Vernon C. Jacobson to be vice president.

Mr. Carson, who farms south of Thompson, was active in the 1950 organization of the Cooperative and became a board member in 1956. He also had served as secretary-treasurer of the board.

Vernon Jacobson, dairy and crop farmer living southeast of Rake, became a board member in 1966.

The final audited costs of plowing in ninety-seven miles of cable for the Grafton and Woden exchanges, which was completed in the summer of 1974, showed total expenses of $168,075.94. This compared with the Cooperative's estimate of $199,381.60. The Cooperative saved some $31,000 by being its own contractor and using its own labor.

At its January, 1975, board meeting, the Co-op's board appointed James Petersburg of Hanlontown to the board to fill the vacancy created by the resignation of Gustaf Gustafson. Mr. Petersburg had recently served for six years as a director of Cooperative Services, Inc., Lake Mills.

Also in January, Manager Lein was elected chairman of the Iowa Telephone Association's Toll and Extended Area Service (EAS) committee. The committee has a major role in operator service agreements and toll ownership for independent telephone companies, and will help solve EAS problems throughout Iowa.

In September, 1975, the Cooperative's board together with Manager Lein decided to exercise the option as negotiated three years earlier, specifically, to transfer its Forest City rural system to General Telephone in exchange for the purchase by the cooperative of the Buffalo Center town exchange.

The decision was generated by the fact that not only were the Forest City rural and Buffalo Center rural the only Cooperative subscribers still without private line service, but General's private line switching terms were unreasonable.

The Cooperative issued a small brochure entitled, "Changes Are Required," and in it gave questions and answers explaining the reasons for exercising the option: increased income for the Cooperative resulting in lower rates overall; improved service for Buffalo Center town and rural subscribers; a $31,000 loss in upgrading switching changes that otherwise would be paid to General Telephone; a loss in toll revenue to General; etc.

As knowledge of the impending system transfers came to the Forest City rural subscribers, an overwhelming wave of resistance to the change was generated. Scores of Forest City rural people appeared at the Cooperative's November board meeting. So large was the crowd that the meeting was moved to the headquarters' garage area. Here the board heard the numerous and persuasive objections and complaints of the group, the thrust of which was "We know we'll get the best service at the lowest rates if we stay with the RTA."

The Board told the group they would meet with them again at a special board meeting in December.

Meanwhile, Manager Lein was instructed and authorized to negotiate with the General Telephone alternatives to the trade plan; namely, new switching equipment and private line service.

At the December 12, 1975, meeting with Forest City rural subscribers, board president Arlo Carson read a long statement, the gist of which was that there would be no trade-off of the Forest City rural subscribers; they would remain with the Cooperative. Here are excerpts from that statement:

"As trustees of the Winnebago Cooperative member, we directors are essentially custodians of the Cooperative's property and are charged with safeguarding and managing Cooperative assets entrusted to us; in setting long-range objectives and making long-range plans; in exploring the economic feasibility of supplying those needs under changing times and conditions. In September, the Board of Directors published an information brochure that discussed alternative methods of up-grading rural Forest City and Buffalo Center areas to one-party lines. In the interest of the whole Cooperative, we believed the trade of properties to be the best economical solution. However, objections from rural Forest City customers were numerous and persuasive while expressions from our other customers in other exchange areas have been minimal. We understand and sympathize with your many objections and it is our unanimous decision to immediately begin renegotiating with General Telephone Company upgrading alternatives that do not include trading properties."

The Forest City Summit newspaper reported that "the exchange of properties between General Telephone Company and the Winnebago Rural Telephone Association 'will not be made at this time,' according to a letter from General Telephone to its customers last week.

"According to Kenneth Lein, manager of the Cooperative, the RTA staff is reviewing with General Telephone alternatives other than the original property exchange plan," the article concluded.

To the extent that Forest City rural was against being traded to General Telephone by the Cooperative, to the same extent were the Buffalo Center town subscribers *in favor of* being traded to the Cooperative. Twenty-three members of the Buffalo Center Community Betterment Club meet April 12, 1976, and formed a committee headed by Wilbert Hassebroek to meet with the Cooperative's board of directors with a view to re-opening the trade question.

The Buffalo Center committee met with the Cooperative's board on April 21, asking "When are we going to receive upgraded one-party service and what has happened to the proposed trade of rural Forest City for the town of Buffalo Center?"

The Cooperative's board reviewed with the visitors the many objections and legal questions raised by the RTA's rural Forest City customers who objected to being traded to General Telephone. The directors and Manager Lein described the alternatives being developed to provide upgraded service to the two rural exchanges without trading properties. "Meetings have been held with General Telephone, the Iowa Commerce Commission, cost consultants and engineering firms," Manager Lein said.

"The cost studies are being developed and further meetings are being planned; everything possible was being done to provide you with one-party service at reasonable rates at the earliest possible date," Manager Lein concluded.

Manager Lein later reported to the board that the Cooperative's revenue requirements to provide its own switching for one-party service to the two rural communities would be approximately $9.37 a month; but that if the Cooperative could negotiate a $9.00 switching fee with General Telephone, such would be the most economical way to proceed.

As it happened, in August, 1977, the Cooperative's board of directors executed a new ten-year switching agreement ($4.25 a month per subscriber) with General Telephone, which would permit the Cooperative to proceed with switching its Forest City and Buffalo Center rural subscribers to all private line, pending the installation by General Telephone of the necessary additions to its switching equipment; and by October, 1978, all of the Buffalo Center rural subscribers were single-party, and a year later in October, 1979, all of Forest City rural were single party.

* * *

E911 Service

At this time, also, Manager Lein reported to the board that several Minnesota legislators were pushing hard to have the emergency 911 phone number service implemented in both urban and rural areas of Minnesota.

Manager Lein pointed out that "the principal 911 advantage as it relates to our area is that even a stranger to the area can reach an emergency number by dialing 911, but who is going to coordinate, answer and dispatch the 911 calls in each of our small communities? It should not be the responsibility of the phone companies to set up a twenty-four hour system; and it is doubtful if any individual small community will accept and carry out that responsibility," Manager Lein said.

On September 24, 1986, Iowa Governor Terry Branstad appointed Manager Lein to a six-year term on an eighteen member State Emergency Telephone Number (911) Commission, which is responsible for recommendations for implementing a 911 program throughout the State.

In 1987, the Minnesota legislature mandated E911 service for all of Minnesota. The Cooperative's Emmons, Twin Lakes, and Conger exchanges provide E911 service ("E" for enhanced) to their phones. "Enhanced" means that the 911 number is answered by an operator with a computer screen, as described below. The number 911 without the E provides for answering service (but no computer) which could be the fire department, or sheriff, police or any public body who could take information and summon required help.

To summon the police, fire department, sheriff, ambulance, rescue squad, or help in any emergency, the caller just dials 911. The signal goes over dedicated trunks to a Public Service Answering Point (PSAP), which for the Cooperative's Minnesota exchanges is in Albert Lea, Minnesota. At the PSAP is a CRT screen, much like a TV screen, which displays the name of the caller, his or her telephone number, address and other pertinent information, such as "invalid in first floor", "children in upstairs bedroom", etc.; this information for the respective residences previously having been gathered and entered into a data base at the PSAP.

A great advantage of the 911 system is the number 911 itself; the caller, be he or she a local resident, a transient worker or other traveler, or any frantic person, can dial "911" without having to look up or remember a seven-digit number. Even if the caller is unable to talk, the CRT displays the caller's address and crucial information to speed a response.

The bulk of the installation and service cost for the Minnesota 911 system comes from the state's general fund; but as required by Minnesota law, the Cooperative collects a ten cent per line surcharge which is credited to the cost of the system. Today, 1990, Minnesota law requires telephone companies to charge eighteen cents a month to each subscriber line.

The Iowa 911 Commission estimated a twenty million dollar price tag for a statewide 911 system. Currently each county has an opportunity to present a plan for state approval on how the county would implement an E911 service. Each community could currently vote a surcharge of not more than twenty-five cents a month for each subscriber for the installation and maintenance of a 911 system; but since it is generally believed that the twenty-five cents will fall far short of actual cost, there is legislation being considered to increase the maximum from twenty-five cents to one dollar a month.

On August 1, 1989, 911 emergency call service was provided for Cerro Gordo County residences, including eighty subscribers in the Co-op's Fertile exchange. All telephone customers in the county will pay a twenty-five cent monthly fee.

* * *

On Sunday, June 29, 1975, Open House was held at the Cooperative's new headquarters building in Lake Mills. More than 600 members and friends toured the building and had the facilities explained to them. Coffee, pop, doughnuts, and cookies were served.

At the Cooperative's annual meeting in September, its Silver Anniversary, Manager Lein reported a savings of $12,202.00 to the Cooperative by doing its own construction in the installation of buried drops and service in the Emmons and Lake Mills exchanges. Estimated cost was $51,482.00; actual was $39,279.00.

Manager Lein reported 1974 revenues of $766,583.05 and expenses of $607,742.01, which included reserves for depreciation and amortization of $155,602.53, leaving $158,841.04 for loan application, dividends and surplus. The REA loan balance stood at $2,501,743.02.

At the meeting, the following retired directors were honored for their dedicated efforts and contributions to the twenty-five years of growth and progress of the Cooperative: Gustaf Gustafson, Forest City, past president; Clarence Willand, Lake Mills; R. E. Aukes, Buffalo Center, past president; M. J. Madson, Northwood; Norman Larson, Clear Lake.

Following the meeting, the board met and re-elected Arlo Carson as President; Vernon C. Jacobson, Vice President; and Victor Knudtson, Secretary-Treasurer.

In November and December of 1975, Manager Lein devoted considerable time in working with the chairman of the Iowa Legislature's House Ways and Means Committee, Al Miller, and with House Member Lowell Norland, Kensett, to ensure that the Iowa telephone cooperative's exemption from rate regulation would continue if the Iowa Commerce Commission were to have jurisdiction over the respective boundaries of the cooperatives.

Briefs - 1975

The Iowa Commerce Commission has ruled that all Iowa telephone companies reduce the maximum number of rural customers on a party line from eight to four, and the maximum number of customers on a town party line to two.

—

The January *Dial Tone* pictured RTA dispatcher Wayne Nyguard relaying a trouble report via radio to the radio equipment field truck of RTA maintenance man Alan Swenson. The RTA has an antenna on top of the Farmers Cooperative Elevator, Lake Mills, giving an effective antenna height of over 200 feet for radio coverage of the entire RTA service area.

—

This winter, Ernest Campidilli and Alan Swenson are teaching an evening class where they instruct other RTA employees in Stromberg Carlson XY switching. Also included in the three hour evening classes is the study of circuit operation from blue prints.

—

According to REA regulations, each member of the Cooperative must be provided with a copy of its by-laws, and this will henceforward be accomplished by printing the by-laws in the Co-op's telephone directory.

—

In August, Ernest Campidilli, Cooperative Operations Engineer, resigned to take a management position in Keosauqua, Iowa.

—

The Woden exchange was cut over to one-party service on March 14, and Grafton was cut over on March 21. The week-long conversions were done on a line by line basis, each subscriber being disconnected from the original multi-party line and reconnected to a dedicated two-wire line.

—

Highway No. 9 from east of Thompson west to Buffalo Center will be widened this summer, and much of the Cooperative's telephone cable currently located thirty to thirty-five feet inside private property fence lines, nevertheless, will fall within the new construction right of way. It will either be lowered to prevent highway construction damage or, when such is not economical because of the smallness of the cable, be replaced with new cable.

—

The June *Dial Tone* notes that this is the twenty-fifth anniversary year of the Cooperative, which is a consolidation of sixty-seven small telephone companies and serves more than 4,800 subscribers in seventeen exchanges in Iowa and Minnesota.

—

Former Cooperative Manager, Glenn Bergland, who is owner/operator of the Lake Mills Telephone Company, is currently president of the Iowa Telephone Association and is a member of the board of directors of OPASTCO—the Organization for the Preservation and Advancement of Small Telephone Companies.

—

Excavating crews cut the 25-pair toll cable south of Rake and the Northwestern toll cable near Garner on June 20 and July 2, respectively, leaving Rake customers without toll service for two and one-half hours in the first instance; and in the second instance customers of Rake, Buffalo Center, Thompson, Crystal Lake, Scarville, Leland, Forest City, Lake Mills, and Woden were without toll service for three and one-half hours.

—

With a vote of seventy-five for and three against, the citizens of Crystal Lake renewed the Cooperative's twenty-five year franchise in that town on November 14.

—

Touch Tone Dialing

In 1975, touch tone dialing was introduced to America. The Cooperative began the conversion from rotary dial to touch tone with the Emmons, Conger, Twin Lakes and Lake Mills (rural) exchanges.

Rotary Dials contain a contact that opens and closes the circuit connecting to the dial office switchboard. After the finger is removed, the dial opens and closes the circuit on its counter-clockwise return to a stopped or normal position. The resulting pulses operate relays in the dial office, which in turn set up a call path for connection to the called party.

Touchtone Dials: Each push button on the dial, when depressed, transmits a dual audio frequency to an electronic receiver in the dial office switch. The tones are then converted to pulses of direct currents of electricity that operate the dial switch, which in turn sets up call paths to the called party.

Touch tone not only is faster, easier and more accurate than rotary dial, but its truly exciting feature is that with it the caller can access various data bases—libraries, banks, and other computerized systems.

Audits

Audits of the Cooperative's books and records are important, just as they are with any business corporation. For the year of 1989, for example, four accountants from the CPA firm of Bobby, Anfinson and Crandall, Austin, Minnesota, worked five days in the offices of the Co-op working with Terry Wegener, the Cooperative's Accountant/Office Supervisor, auditing the Cooperative's books. Their written report of the audit was given to the board of directors for their review, and a report was compiled for the annual meeting.

Caution. Men at work. This photo of a number of years back shows Alan Swenson, upper left, Marlin Senne, lower left, and Darwin Adams and Wayne Nyguard on the right, left to right, repairing underground cable damage.

1976

Cooperative Was Being Inadvertently Shortchanged

Early in 1976, Manager Lein and consulting engineer Dale Jacobson made a surprising discovery: The Cooperative was not receiving its proper share of certain long distance revenues from Northwestern Bell Telephone!

Manager Lein and Dale had gone to the Grinnell, Iowa, headquarters of General Telephone to check General's books and records as they related to toll calls from the Cooperative's Forest City and Buffalo Center rural subscribers. They discovered that a portion of the long distance fees was being retained in the Bell toll pool.

There followed by the Cooperative numerous attempts to obtain a proper division of toll, but without success; so the Cooperative began to withhold money at the rate of $4,000 a month from its monthly toll settlement with Bell.

In September, 1977, Northwestern Bell filed suit against the Cooperative to recover the withheld funds. The Cooperative immediately filed a counter suit. It hired an attorney from Topeka, Kansas, namely, James Caplinger, a communication's attorney; and he, together with the Cooperative's attorney, Dudley Weible of Forest City, pursued the Cooperative's claim in conference with Manager Lein.

Numerous meetings, interrogatories, and depositions were conducted; and in November, 1979, Northwestern Bell filed a motion for summary judgement in the district court in an effort to get the case dismissed in Northwestern's favor.

A hearing was held on January 7, 1980, and shortly thereafter, the district judge denied Northwestern Bell's motion and set a trial date for February 26 at Forest City. At the same time, the judge instructed both parties to the suit to try and settle out of court.

Although previous attempts had been made, an out of court settlement was pursued again; and after several meetings, Northwestern Bell agreed to an out of court settlement in which the Cooperative would receive $150,000 in back toll revenue generated from the two rural exchanges of Forest City and Buffalo Center.

Furthermore, Northwestern Bell would pay the Cooperative one-half of an *average schedule* payment for tolls generated by the two exchanges. Such payment comes to approximately $5,000 a month—but not less than $4,000—and compensates the Cooperative for its rural plant toll investment.

Thus a situation was corrected in which Northwestern Bell's toll pool was being overcompensated as a result of the complex switching arrangements between Bell, General, and the Cooperative.

(Editor's note: Again, we have evidence of the excellent stewardship of the Cooperative's affairs given by the management; first by Manager Bergland and now by his successor, Manager Lein.)

After the 1984 AT&T divestiture, the rules were changed and the matter again came under consideration by the Cooperative. In 1990, the Cooperative cancelled its switching agreement with General and installed its own separate, computerized digital dial switching equipment in Forest City and Buffalo Center to switch the Co-op's rural subscribers in the two communities.

* * *

The announcement of the settlement appeared in the January, 1990, *Dial Tone*: "The Cooperative's telephone customers in rural Buffalo Center and Forest City since 1950 have received dialtone and telephone services from switches owned by GTE in the downtown areas of Buffalo Center and Forest City. "Although this arrangement has worked well over the years, if rural Buffalo Center and Forest City customers are to receive the same types of service that other RTA customers receive, the current switch arrangements must change.

"The new installation will allow the Cooperative to offer a choice of competitive long distance carriers to customers in rural Forest City and Buffalo Center. Their long distance calls will be routed through the Lake Mills switch where they will be recorded, rated, and delivered to the fiber optic facilities of Iowa Network Services' toll tandem switch in Des Moines and to the world.

"Custom calling features will include call forwarding, call waiting, three-way calling and speed calling. We point out that these state-of-the-art phone features are not yet available in countless phone systems in the country."

RTA crews installed buried drops and service wires in the Conger and Twin Lakes areas during the summer, replacing the aerial drops and wires to the farm homes. Two-pair wires were used in the individual connections in anticipation of a customer's requesting a second telephone line. The service wires cables, laid thirty inches below ground, are filled with petroleum jelly to keep out moisture and have an outside five-mil-thick bronze shield.

In August, 1976, Manager Lein observed his twenty-fifth anniversary of employment with the Cooperative. "Since I began work in August, 1951, the technical changes and improvements in electronic equipment have occurred at an ever increasing rate," Manager Lein said. "These new services have made the years both challenging and deeply rewarding to me," Lein added.

The Cooperative's annual meeting was held on September 16, 1976, with nearly 1,000 members and guests in attendance. Manager Lein reported 1975 revenues of $836,642.91 and operating expenses of $492,790.60, leaving $164,852.31 for depreciation and amortization expense, and $179,057.17 for REA loan repayments, dividends and surplus.

Working manager and tool of the trade—his telephone. Ken Lein. (1978)

The careful, frugal management of the Cooperative was reflected in a "cash on hand and in banks" balance of just $6,652.83 and $430,538.05 invested in interest-bearing bonds.

At the meeting, the members voted to remove the three three-year term limitation for board members.

Incumbent directors, Arlo Carson, Milford Helgeson and Lloyd Kronemann were re-elected by vote of the members. Following the meeting, the board convened and elected Vernon Jacobson, Rake, as President; Victor Knudtson, Lake Mills, Vice-President; and Lloyd Kroneman, Grafton, Secretary-treasurer.

On October 10, 1976, the Cooperative installed its five thousandth main telephone in the residence of Mr. and Mrs. Philip Davis, Route one, Grafton. The Davises were awarded a ten dollar toll credit certificate, and a bronze plaque was affixed to their telephone which read: "5000th Telephone, Installed October 20, 1976." Mr. and Mrs. Davis and RTA employee Daryl Stelter were pictured in the January, 1977 *Dial Tone*.

Mr. and Mrs. Phil Davis, Grafton, and 5,000th RTA telephone.

On November 29, 1976, the Lake Mills Telephone Company completed its old dial central switching conversion to new electronic dial equipment, and between midnight and 6:00 a.m. November 30, all Lake Mills town and rural customers were connected to the new switching center. New optional services made available included push button calling; call forwarding; call transfers within the Lake Mills and extended area exchanges (EAS); speed dialing—using one or two digits only; and call waiting—a busy line beep alert for incoming calls.

The Cooperative had anticipated the new service would bring increased usage of telephone service in the Lake Mills area and had installed additional EAS trunks from the Leland, Joice and Scarville exchanges.

Briefs - 1976

On January 6, Eugene Walz began work with the Cooperative. After high school, Eugene attended Wadena Area Vocational Technical School, Wadena, Minnesota, studying electricity, electronics, mathematics and telephone installation.

—

Foster Hartman, this February, was promoted to the position of Operations Manager for the Cooperative.

—

The Cooperative will take applications for summer construction work "from local persons who have completed at least one year of college and who plan on returning to college in the fall," the March *Dial Tone* reported.

—

Ron Zielke, owner/manager of Ron's Spray Center, Twin Lakes, Minnesota, was pictured in the March *Dial Tone* "enjoying his recently installed WATS telephone service."

—

The increase in postage rates to thirteen cents from ten cents will increase the Cooperative's mailing costs this year by $2,500.00, to more than $12,000.

—

The cover of this year's Directory depicts the painting, "Summer Marsh," courtesy of the artist, Viola Buckley, Fertile, Iowa.

This RTA employee guides a rotating pipe that is drilling a hole for telephone cable beneath a road. The device in his hand is a motorized right-angle chuck that receives its power supply from a 240-volt portable generator, out of sight in the picture, via the power line leading from the chuck. Connected to the pipe is a water hose, lower center, that supplies water for facilitating the drilling and cleaning the hole. Hip boots are a must.

1977

In late 1976, the Fourth Circuit Court of Appeals in Richmond, Virginia, suspended a ruling of the Federal Communications Commission that would have permitted telephone customers to attach their own equipment to telephone company lines without a protective device. The appellate court ruling in effect left the jurisdiction over such installations to the individual state regulatory commissions.

The telephone industry opposed the FCC ruling on the grounds that customers' complete control of their telephone equipment would be costly and cause serious problems.

In February, 1977, there was agitation by some legislators and a couple of members of the Iowa Commerce Commission to rate re-regulate the telephone industry. Maurice Van Nostrum, chairman of the Iowa Commerce Commission, however, did not believe that small phone cooperatives and mutuals should be rate regulated; their members are owners and able to control the rates which they, as users, want as low as practicable. Former Manager Bergland and Manager Lein were instrumental in composing an audio tape and slide presentation for showing throughout the state to explain the logic of the small Telcos' exemption from rate regulation.

The Cooperative continued its winter night classes for employees. Twelve three-hour sessions were held in the RTA Lake Mills garage. The instructor was Foster Hartman and class members included Daryl Stelter, Eugene Walz, Marlin Senne, Alan Swenson and Norman Risius. Subjects covered included testing, installation and maintenance of subscriber and trunk electronic radio carrier equipment. "Trunk electronic radio carrier" systems operate over existing cable pairs where such is the most economical method of adding additional circuits. "We are proud and pleased that Cooperative employees are willing to take evenings of their own time for self-improvement courses so as to provide better service for the cooperative's members," Manager Lein said.

One thousand members and their families attended the Cooperative's annual meeting on September 22, 1977. Manager Lein reported 1976 revenues of $906,812.83 and expenses of $740,222.87, which included $183,196.13 reserved for depreciation and amortization expense, leaving $166,196.13 for loan payments, dividends and surplus. The REA loan balance as of December 31, 1976, stood at $2,306,917.57.

The members re-elected incumbent directors Vernon Jacobson and DeVere Steffensen to three-year terms.

Following the meeting, the board of directors convened and elected Vernon Jacobson as Board President; Victor Knudtson, Vice President; and Lloyd Kroneman, Secretary-Treasurer.

The October, 1977, *Dial Tone* reported that "toll calls placed by RTA customers use the lines and equipment of the Cooperative and at the same time, the calls are extended over facilities usually owned by the Bell Telephone Company. The toll revenues received from each toll call are divided between independent companies and the Bell Company on an 'average settlement basis.' This average settlement formula is worked out between the United States Independent Telephone Association, of which your cooperative is a member, and the American Telephone and Telegraph System of which Northwest Bell is a subsidiary. The settlement is worked out and agreed to by all parties involved."

On November 1, 1977, Manager Lein was elected a director of the Iowa Telephone Association, whose membership comprises 167 Iowa independent telephone companies and the Bell System.

RTA Members Receive $104,500

Checks totaling $104,519.33 are signed in December, 1977, by Vernon Jacobson, Ken Lein and Lloyd Kronemann, respectively president, manager and secretary-treasurer of the Cooperative. The checks, which went to the Cooperative's members, represented dividends previously deferred for the years 1950 (when the Cooperative was organized) through 1961.

In December, 1977, the Cooperative paid its first deferred patronage dividend by distributing $104,500 to the members. The amount paid to each member was proportioned to service, local and long distance, purchased by the member in the time period from 1950 when the Cooperative was organized through 1961.

Called capital credits, such dividend payments are unique to cooperatives. "Margins, created by excess revenue over costs, are created by you, the member-customers, to build financial strength in your Cooperative," Manager Lein said.

"Unlike investor owned utilities, which issue stock, margins comprise the only source of equity capital or member ownership in cooperatives.

"Between the time margins begin to accumulate and until they are returned to members, they are used to finance the Cooperative's construction, improvements and operations.

"This equity provided by you represents your investment in the system, and although you receive no return on your equity investment, your Board of Directors feels that each member deserves an ultimate return on your investment.

"Your Cooperative has achieved a financial condition that will allow the payment in cash of some previously deferred dividends. Dividends for each year in which allocations were made from the Cooperative's beginning through the year 1961 will be paid.

"The revolving of capital on a first in-first out basis is the very foundation of cooperative operations. However, members must realize that funds to retire patronage capital invested by members in prior years must, without exception, come from patrons served now and in the future.

"To date, very few telephone cooperatives have begun capital credit retirements; your Cooperative is unusual in this regard. Not only has your Cooperative been able to provide excellent telephone service on a day-to-day basis at low rates, but it has accumulated sufficient reserves to complete one-party construction and begin to pay deferred dividends in cash, "Manager Lein concluded.

In late fall of 1977, the Federal Communications Commission (FCC) issued rules governing ownership by customers of their own telephone sets. Included were the following provisions, mandated by law:

Customer owned sets may be connected only to single party lines; all connections are to be made through standard jacks provided by the telephone company; before making a direct connection, the customer must notify the telephone company of terminal equipment he plans to connect and to which line; customers must provide the telephone company access to customers' premises to facilitate testing and maintenance of telephone company owned wiring and jacks; the telephone company shall provide wiring and standard jacks as per filed tariff, keep records of customer owned installations, and provide technical information on ringers, frequency, etc.

Telephone company will not: Install customer provided equipment, modify customer provided equipment to make it compatible with telephone system, check customer's installation equipment for performance, or provide maintenance on customer provided plug or terminal equipment.

Briefs - 1977

In February, 1977, the Cooperative advertised the availability in the Conger, Emmons and Twin Lakes exchanges of touch tone dialing at $1.50 a month per line phone and fifty cents a month for each extension phone.

—

In 1977, just seven states—Minnesota, Iowa, Missouri, North and South Dakota, Nebraska and Kansas—accounted for almost half (43%) of all the American farm cooperatives.

—

A Pioneer: In 1950, when the Winnebago RTA got its first REA telephone loan, one of the first in the nation, only thirty-eight percent of U.S. farms had phones! Today ninety percent have phones.

—

The June *Dial Tone* pictures Mrs. Gary Honsey, Emmons, using her new touch tone wall phone.

—

A good telephone neighbor and friend of the Cooperative passed away in October: Arthur Dietrich, manager of the Titonka Telephone Company.

—

The cover of this year's Directory depicts the painting, "Geese at Rice Lake," courtesy of the artist, Floyd Thomas, Lake Mills, Iowa.

1978

By now, the Cooperative had installed touch tone service capability in all its exchanges except Buffalo Center rural and Forest City rural, where the switching equipment was owned and controlled by General Telephone.

Manager Lein announced a savings of $17,000 in construction expense for burying the drops and service to the farmers in Rake, Scarville, Joice, Leland, and Thompson areas. Construction cost estimate was $107,000; actual cost was $90,000 by using the Cooperative's own crews for the project.

Grafton: 100 Years

Grafton celebrated its one hundredth anniversary in 1978. This community, with its population of largely German descent in its early years and still today, has won awards for being an exemplary small town, in Iowa and nationally. Grafton's history, 1878 through 1978, was published in book form, eight and one-half by eleven, 164 pages; and on page eighty-eight is the history of the Grafton Telephone Company, which was consolidated into the RTA Cooperative in 1961. It was originally organized under the name of German Telephone Company—Deutsche Fermsprecher Kompanie.

During the early years, telephone member dues were $2.00 a year. A man and team were hired in 1922 to set new telephone replacing broken ones at $3.00 for an eight hour day. Operators were paid $30.00 a month and were on duty sixteen hours a day except Sundays and winter time, when the hours were shortened.

Henry Tesch was the first Secretary-Treasurer of the phone company, followed by William Dalluge, Albert Boettcher, C. J. Verchow, Herman Miller, William Bublitz, L.A. Murphy and Walter Wagner (1961).

Repairmen who served during the company's existence included Earl Miller, Herman Miller, Lavern White, Sidney Homes, Lowell Schultz and David Diedrich.

Grafton telephone operators through the years included Alma, Julie and Myrtle Ueker, who were first operators because the first switchboard was in the August Ueker home; others, but not in chronological order, included Violet Dalluge, Pearl Virchow, Wilma Macken, Vinora Virchow, Dena Gettschalk, Alice Rosenberg, Edna Virchow, Clara Bork, Dora Schultz, Elsie Schmidt, Mildred Schauer, Adeline Bublitz, Lydia Borchardt, Erma Urbatsch, Leona Wilde, Ivy Kirchgatter, Arminda Buechele, Lola Ueker, Leatha Trettin, Della Trettin, Myrtle Feldt, Marcella Schmidt, Marilyn Schmidt, Luetta Schmidt, Virginia Ueker, Harriet Christians, Hazel Schultz, Lizzie Kirchgatter.

* * *

Early in 1978, the Cooperative applied for a $1,400,000 REA loan to upgrade Buffalo Center and Forest City rural to one-party service, to provide service to an anticipated 500 new subscribers in the next five years, and to add touch tone, modular jack wiring, toll and extended area service additions. (Buffalo Center rural became all single party service in September, 1978.)

In 1978, also, the Cooperative established a phone center in its central office, with a large inventory of different styles and colors of phones for purchase by customers as their own phone equipment.

Local Phone Service for Less Than It Costs

At the Cooperative's annual meeting in September, 1978, Manager Lein explained how the charge to customers for local telephone service is less than the actual cost of providing that service by the Cooperative. "The difference between the cost of providing local service and what you pay for it is made up by long distance revenues," Manager Lein said.

"Indeed, nationally, out of every two dollars of long distance revenue, about fifty-five cents helps to pay the cost of local telephone service.

"Why do we underprice home telephone service?

"In America, telephone service is based on the idea that every family should have reliable telephone service at a reasonable cost. America's telephone companies, with government approval, have pretty well made that concept a reality.

"We think using long distance revenues to help keep charges for local telephone service down so virtually every family can afford it is more than just a good idea. It's a national policy that has served our customers and the country well for years," Manager Lein concluded.

At the meeting, Manager Lein reported Cooperative revenues for 1977 of $951,489.02 and expenses of $788,675.82, which included reserves for depreciation and amortization of $188,550.97, leaving $162,813.20 for REA debt repayment, dividends and surplus.

Members re-elected incumbent board members Victor Knudtson and James Petersburg to new three-year terms.

Following the meeting, the board convened to re-elect Vernon C. Jacobson as President; Victor Knudtson, Vice President; and Lloyd Kroneman, Secretary-Treasurer.

Briefs - 1978

On January 16, Julie Wakeman began work in the Cooperative's general office. "We look forward to the contribution that Mrs. Wakeman will make to the Cooperative," Manager Lein said.

—

Colonel Piersol has completed twenty-five years of service with the Cooperative. "All of us at the Cooperative, employees and directors, thank Mr. Piersol for those twenty-five years, during which he has always given his best," Manager Lein said.

The RTA's second retiree, Colonel Piersol, right, accepts gift from Manager Lein at retirement party, December, 1978. Colonel Piersol, service technician, joined the Cooperative in 1953. Glenn Bergland was the Cooperative's first retiree.

—

In 1940, six hours average pay paid for one month's telephone service. Today, two hours' pay does.

—

RTA employee Daryl Stelter attended a switch school thirty-six hour class in Houston, Minnesota, taught by a representative of the Stromberg-Carlson Company.

—

The July *Dial Tone* pictures RTA's Alan Swenson testing additional new extended area circuits between Joice and Forest City, Joice and Kensett, and Joice and Hanlontown.

—

On September 5, 1978, Jay Hagen began work for the Cooperative. A graduate of Lake Mills High School and Jackson, Minnesota, Voc-Tech, Jay was employed by Finley Engineering firm in Iowa City.

—

Connie Harringa, daughter of the Melvin Harringas southwest of Buffalo Center, is pictured in the October *Dial Tone* using the family's new touch tone phone.

—

The town of Grafton was the scene for a television commercial, shot by the St. Regis Paper Company, that compared the volume of the residents' weekly garbage bags with the same amount of garbage when compacted by a garbage compactor into St. Regis laminated garbage bags. Each of the town's 120 families was given a $160.00 garbage compactor by the paper company, which selected Grafton because of Grafton's appearance on several "Best City" lists.

—

Anthony Coloff and his wife, Sue, have gone on the air with their new FM radio station, KIOW, Forest City.

Here is the RTA's switch building at Woden, showing the emergency 3000-watt standby generator mounted outside. In the event of a commercial power failure, the generator automatically takes over. Actually, the dial switchboard in the building is powered by a set of batteries, which are charged by the commercial (or generator) power. The batteries serve to filter extraneous noise from telephone transmission, which otherwise would prevail if the switches received their power directly.

1979

Early in 1979, Minnesota de-regulated telephone rates for small companies, following the guide established by Iowa. "It's been clearly proven that rate de-regulation for small companies has resulted in a lot of telephone companies being able to survive in Iowa. If anyone today compares the profitability and the low rates of Iowa telephone companies with any of the phone companies across the nation, he'll find that rate de-regulation in Iowa has been good for the customers and the companies," Manager Lein observed.

In January, 1979, the Cooperative received REA approval for an REA loan of $1,425,900. To this was added $149,000 of the Cooperative's own

This 1979 photo of directors and past presidents of the Iowa Telephone Association shows director Lein, standing, fifth from left, and president Bergland, seated, third from right.

funds, and the total used to rebuild the Forest City rural exchange to single party service, provide initial service to 420 new subscribers, install 1,160 new extensions, add central office lines to all exchanges, install toll carrier in seven exchanges, install tone dial instruments and gas tube lightning arrestors, six extended area carrier systems, and TSPS in eleven Iowa exchanges. TSPS is Traffic Service Position System, which allows the caller to dial zero followed by the called number to reach the operator to charge the call, as opposed to formerly having to dial zero and wait for the operator to take credit instructions and complete the call.

The $1,400,000 REA loan will bear interest at seven percent per annum.

As mentioned earlier, loan approval does not mean that the Cooperative receives a check for the full amount of the loan. Instead, the loan amount is set up by the REA as a credit to the Cooperative and funds against the credit are received piecemeal by the Cooperative in response to REA approved requisitions from the Cooperative for the specific completed projects listed in the loan application.

Manager Lein, in January 1979, was elected to the board of directors of OPASTCO, the national Organization for the Protection and Advancement of Small Telephone Companies, with a membership of more than 300 rural independent phone companies serving more than one and one-half million customers in 1,175 exchanges coast to coast.

The monthly publication *General News*, of the General Telephone Company of the Midwest, reprinted from the June, 1979, *Dial Tone*, an essay by Manager Lein on telephone cooperatives.

The thrust of Manager Lein's message was that rural telephone systems are different from large urban systems and "there is no way rural America can overcome the handicap of low density and deliver comparable services (as compared to urban) at comparable prices absent mechanisms to bring dollars to low density areas," wrote Manager Lein. His message was in effect directed to the Congress, asking that body to take legislative action reflecting the true interest of rural telephone companies, in response to recent decisions of the Federal Communications Commission favoring competition in the providing of communications equipment and intercity private line service.

"In low density areas, it is essential that regulatory policy recognize the area-wide nature of telecommunication service offerings. Companies and industries who offer the same technology but do not offer that technology to all consumers within the area are not competitive," Manager Lein wrote, and that includes "CATV (cable TV) companies," he added.

The 1979 annual meeting was September 20 at the Lake Mills High School gymnasium. Manager Lein reported 1978 revenues of $1,055,704.55 and expenses of $870,033.41, including $197,779.35 reserved for depreciation and amortization, leaving $185,671.14 for loan payments, dividends and surplus. The Cooperative's REA debt stood a$1,060,499 on December 31, 1978.

In his remarks to the members, Manager Lein foresaw phone services that have since become commonplace, at least with the Winnebago RTA: "Special calling features can now be installed in even relatively small central offices, which provide call forwarding, conference calling, call waiting and speed dialing," Manager Lein said. "Facsimile transmission is now possible . . . rural companies must be allowed . . . to provide a full range of broad band services such as cable TV."

Arlo Carson, rural Thompson, retired from the Cooperative's board of directors after twenty-three years of service on the board, effective at the close of the annual meeting. Mr. Carson was presented with a twenty-three year service award at the meeting. "Since 1956, when he came on the board, Mr. Carson has been a vigorous proponent of improved telephone service, all single party lines, and services that are beneficial to rural telephone customers," Manager Lein said. "During his twenty-three years of service, the Cooperative grew from eight exchanges to seventeen, from 3,000 customers to over 5,000, from assets of one million dollars to assets over four million dollars, and it continues to be a model telephone cooperative in the REA program. A sincere thanks to Arlo Carson and for the time and effort he has given to the Cooperative," Manager Lein said.

Keith Finer was elected to replace Mr. Carson. Keith and his family operate a grain and dairy farm southwest of Thompson. He and his wife, Beverly, are the parents of five sons.

Following the meeting, the board convened to re-elect Vernon Jacobson as President, Victor Knudtson, Vice President, and Lloyd Kronemann, Secretary-Treasurer.

Briefs - 1979

The February *Dial Tone* pictures Wayne Roberts doing repair work on the Cooperative's HD20 Allis Chalmers, which was still going strong since purchased used in 1968.

—

RTA employees Jay Hagen and Gene Walz received graduation certificates from Manager Lein for completing an electronics course at North Iowa Area Community College, Mason City.

—

Neal Sletten began employment with the Cooperative this December. After graduation from Lake Mills High School, Neal was employed by Winnebago Industries in Forest City, and Larson Manufacturing in Lake Mills. Neal was assigned to the RTA's construction and outside maintenance department.

—

Former Manager Glenn Bergland is featured in the twenty-five year history book of the National Cooperative Telephone Association, 1954-1979. Manager Bergland was a NTCA founding board member, serving from 1954 through 1961 as representative of Region Five of NTCA's National Regions. He served as NTCA President from 1959 to 1961.

—

The cover of this year's Directory features the photo of an old shed in the midst of a riot of autumn woodland color, courtesy of the photographer, John Carlson, Lake Mills, Iowa.

HD 20 crawler tractor, center, is pulling cable plow, headed across a narrow gravel road.

1980

Effective January 1, the federal excise tax on telephone service was reduced from three percent to two percent, which translated into an annual savings of $14,000 for Cooperative members. The tax, which was first imposed during World War II, has ranged as high as fifteen percent on local services and twenty-five percent on long distance. The tax reduced to one percent on January 1, 1982.

Manager Lein presented graduation diplomas to the following RTA employees for completing ten evening classes in electrical and/or advanced electronic courses at NIACC: Mark Schultz, Gene Walz, Daryl Stelter, Neal Sletten, Jay Hagen, and Norman Risius.

Mr. and Mrs. Ken Lein and Mr. and Mrs. Glenn Bergland at the Iowa Telephone Association annual meeting. (1980)

The Cooperative announced plans to convert its Iowa exchanges to Direct Dial service beginning in June, 1981, with new toll facilities to be both buried underground and installed as electronic radio carrier circuits. All fourteen Iowa exchanges were converted by February, 1982.

The 1980 annual meeting was September 17 in the Lake Mills High School gymnasium.

In speaking to the members, Manager Lein expressed concern over the Internal Revenue decision that in order for cooperatives to maintain their non-profit IRS tax exemption, no more than fifteen percent of their income can come from co-op non-members. "In recent years, Manager Lein said, the IRS has taxed telephone directory advertising, rental agreements and yellow page ads. The IRS decision is a critical one." (The fifteen percent limitation is still in effect today, 1990.)

Manager Lein reported 1979 revenues of $1,189,298.03 and expenses of $928,161.72, including $221,597.68 reserved for depreciation and amortization. This left $261,136.31 for loan repayments, dividends and surplus. The Cooperative's REA debt stood at $2,047,227.38.

Incumbent board members Vernon Jacobson and DeVere Steffensen were re-elected by the Cooperative's members to three-year terms on the board.

The board re-elected Vernon Jacobson as President; Victor Knudtson, Vice President; and Lloyd Kronemann, Secretary-Treasurer.

In December, 1980, the Cooperative installed a radio pocket pager system antenna on the REA tower south of Thompson. Pocket pagers relay a beep signal and voice of the caller to the person wearing the pagers. The person receiving the call will answer the page by calling the originating number from the nearest phone. Page range is fifteen to twenty miles. The system requires a license from the FCC. Pager rates are $20.00 a month for company owned pagers and $12.00 for subscriber owned.

The Cooperative installed a one hundred line addition to the Twin Lakes exchange using rebuilt used equipment. "Costs of new equipment have risen enormously in the past few years, and we find rebuilt equipment from a reliable re-manufacturer gives excellent service," Manager Lein said. RTA's Norman Risius completed the installation.

Glenn Bergland was presented a Founders' Plaque by U.S. Assistant Secretary of Agriculture, Alex Mercure, in October for recognition of Glenn's nine years of service on the board of directors of the Rural Telephone Bank. Rural Telephone Bank is an agency of the Federal Government that provides financing to rural telephone cooperatives (see page 113).

In presenting the plaque, Mercure pointed out, "Glenn Bergland has been directly involved in the rural utility industry since 1939, when he became the original manager of the Winnebago Rural Electric Cooperative Association.

"He took the leadership in organizing the Winnebago Cooperative Telephone Association by merging sixty-seven small rural telephone companies in North Central Iowa and Southern Minnesota. The Winnebago RTA received the first loan made in Iowa by the Federal Rural Electrification Administration to a rural telephone cooperative system."

Briefs - 1980

The Rural Electrification Administration in Washington has celebrated its thirtieth year of providing financing to America's rural telephone systems. Over 4.7 billion dollars has been loaned and ninety-three percent of the nation's farms have phones.

—

Every time a hunter shot a duck, his dog would retrieve it by running along the top of the water. The dog did this several times. The hunter said proudly to the onlooker, "Did you notice anything unusual about my dog?" The onlooker replied, "Sure did, he can't swim."

—

John Hanna began employment with the Cooperative on March 31 on the RTA's construction crew.

—

Todd Ranes began employment with the Cooperative on May 27, 1980, after completing the telephone training course at the Jackson, Minnesota, Voc-Tech School.

—

Manager Lein met in Washington, D.C., this May with Senator Roger Jepsen to discuss the revision of the Communications Act of 1934.

—

Did it fly? At the 1980 Annual Meeting, Manager Lein said that Zenith TV offers a new feature called the space phone. When a phone call is received, the TV sound drops out while the picture remains. The caller's voice is heard through the TV set, while a microphone picks up voices in the room for transmission to the caller at the distant end.

—

Speaking of flying, United Air Lines and Page America Communications are testing Sky-Tel, an in-flight telephone call system. (In-flight calls are a common occurrence today, 1990.)

1981

The Cooperative's single party rates have remained the same since conversion from multi-party service; but across the nation in 1981, 468 million dollars in in-rate increases were pending before state regulatory agencies. "Through the dedication of loyal employees, proper planning through the years by a board of directors who understands the telephone system, and by continued effort on everyone's part to keep costs down, we have not had to increase rates," Manager Lein said.

In February, Manager Lein was elected Second Vice President of OPASTCO. The organization has established an office and staff in Washington, D.C., to lobby in Congress "for legislation that will protect against excessive local rate increases if supporting long distance revenue is reduced, and legislation that will ensure against rural long distance users paying a disproportionate share of long distance costs, and guarantee that rural phone companies will be permitted to offer a full range of broadband and other sophisticated telephone services; such as, cable TV, data and computer services.," Manager Lein said.

In April, 1981, *Dial Tone,* Manager Lein expressed concern over President Reagan's proposed cutback in the REA lending program of 187 million dollars in 1981, and 350 million dollars in 1982, by eliminating loans to telephone companies and cooperatives and by reducing loans to electric cooperatives. "The Administration is attempting to justify the cuts on the basis that since telephone service already has been extended to most of rural America, future expenditures should be paid by the customer," Manager Lein said.

"Telephone cooperatives like ours are denied loans by private money markets, which are not interested in capital intensive, low profit, low density, high cost rural telephone systems. Furthermore, we must prove to Congress that the job is not yet done; and new coaxial and fiber optic facilities capable of handling computer, data base, cable TV, electronic newspaper, mail and other broad band services, must be offered to our rural areas. Telephone cooperatives are joining together to show the tremendous economic benefit that has accrued for the whole nation when 'plain old telephone service' was brought to the rural areas." Manager Lein concluded.

At the Iowa level, Governor Ray proposed a three percent tax on interstate long distance calls. Manager Lein pointed out that such a tax not only might be unconstitutional; but in addition, would (a) put Iowa businesses at a disadvantage when competing with tax-free states, (b) would be discriminatory because not applicable to airlines, truck and other common carriers, (c) would make Iowa the only state having such a tax, (d) would discriminate in favor of Iowa toll calls charged on credit cards issued out of state and, hence, tax free, and (e) would unfairly burden RTA customers near the Minnesota border who make frequent calls to Minnesota. "Our efforts so far have prevented the governor's tax proposal from being voted out of committee," Manager Lein said.

The Co-op's Al Swenson repairs a switch part. (1978)

In June, 1981, the Cooperative became a stock holding member of U.S. Intelco, a corporation headquartered in Olympia, Washington, to provide independent telephone companies with services that include storage and retrieval of subscriber data. The Intelco data base is linked to the AT&T operations network and, thus, Intelco members can participate in the various calling services offered by AT&T. Intelco also provides billing and collecting services and generates revenue for its members from the sale of telecommunication industry products and services. Today, 1990, Intelco is owned by 300 independents. Its services are used by some 700 U.S independent telephone companies. Intelco has 135 employees. Currently under construction is a four and one-half million dollar headquarters building in Olympia, Washington.

"Intelco is an exciting project," Manager Lein says. Manager Lein was appointed to the board of directors of U.S. Intelco in January, 1987, and re-elected in January, 1990.

"I believe the future of independence lies in becoming enhanced service providers, using data bases such as U.S. Intelco to meet the long-term communication needs of rural telephone subscribers," Manager Lein said.

"The new network being planned by the industry is called an SS7 network. It is a dedicated digital network quite unlike the conventional telephone switching. It carries information over data links separate from voice path. While the voice transmission continues over existing voice trunks, SS7 is in fact a digital data overlay of the existing network. Using signaling nodes connected by data links, this new network will store information on how to route and process calls while routing individual calls to various centralized data bases.

"I dislike the possibility that large telephone companies would be the SS7 provider for Winnebago, and I'm reluctant to put Winnebago's customer data base information in a 800 data base owned and controlled by the Bell Companies. Control of the subscriber data base is the key to the future of SS7 services; the only way we can ensure that our customers have the right service at the right price.

"Since relying on the Bell Companies does not appear to hold the key to our future, Winnebago, along with a lot of other independent telephone companies, has formed ITN, Independent Telecommunication Network, Inc., to install, operate and maintain an independent SS7 network. U.S. Intelco has given the ITN SS7 network assistance during start-up period; assembling the physical network and currently testing two interconnected transfer points that have been installed at Mattoon, Illinois, and Rock Hill, South Carolina.

We are optimistic about ITN's potential to play a major role in the evolution of the data based intelligent network today and in the future," Manager Lein said.

The Cooperative's Annual Meeting was held September 23, 1981. In his remarks to the members, Manager Lein said the telephone organizations of which the Cooperative is a member have in recent years worked to formulate and ensure the passage of legislation that will ensure "that long distance calls over competitor's lines that use our local exchange facilities will contribute to support the cost of the local exchange; that uniform long distance rates will continue so that the cost of toll calls between two small Iowa towns over a small trunk group or between two large cities are equal; and that small telephone companies must be allowed to enter new markets and find new sources of revenue. This means allowing us to provide data, cable TV and other broad band services to our subscribers."

Manager Lein also said that "the FCC recently ordered the deregulation of terminal equipment, such as telephones, by March 1, 1982. They have also ordered that beginning October 1st this year, we stop capitalizing and begin expensing new station inside wiring which consists of all wire and material from the house protector to the house telephone. Not to be outdone, the Minnesota Public Service Commission issued an order that expensing of new inside wiring be phased in over a four-year period."

Manager Lein told the members that the historical, friendly relationship the Cooperative has enjoyed with the Bell long distance network is deteriorating because of certain FCC decisions.

"Our relationship with Bell has become more competitive and less of partnership," Manager Lein said. "It may be necessary for cooperatives like Winnebago to form consortiums with other small independents to provide our own long distance network over microwave and satellite systems. At the least, we may want to arrange for alternate long distance routes besides the one we now have with the Bell system.

"Earlier this summer, a small telephone company, The Northwest Iowa Telephone Company, made history by reaching an agreement with Microwave Communications, Inc., or MCI, to interconnect to MCI's new long distance network. Customers of Northwest now have the choice of dialing 'one' to place calls over the Bell lines or dialing 'six' to place calls over the MCI lines."

Manager Lein reported that the Cooperative had 1980 revenues of $1,478,104.13, and expenses of $1,142,880.08, including a $292,434.80 reserve for depreciation and amortization, leaving $335,224.05 for debt repayment, dividends and surplus.

From 1950 through 1980, the Cooperative borrowed a total of $4,177,361.13 including interest from the REA, and made principal and interest payments totaling $2,139,297.25, which left net debt, as of December 31, 1980, of $2,038,063.88.

The members re-elected incumbent board members Victor Knudtson and James Petersburg to three-year terms. The board re-elected Vernon Jacobson as President; Victor Knudtson, Vice-President; and Lloyd Kronemann, Secretary-Treasurer.

Robert Houtkooper Fatally Stricken

Robert M. Houtkooper, Accountant/Office Supervisor for the Cooperative, died suddenly on October 16, 1981. Mr. Houtkooper had worked for the Cooperative since 1973. "Bob was a fine employee, very capable and very loyal; and all of us in the RTA will miss Bob's quiet, friendly presence," Manager Lein said.

Waiver for Lake Mills Cable TV

In November, 1981, the FCC eliminated the telephone/cable TV cross-ownership rule as it related to communities of less than 2,500 population that are not part of an urbanized area. The "cross-ownership rule" forbids a telephone company from owning a cable TV system in a community of more than 2,500 population. This rule change allowed the later installation of the RTA's Thompson cable TV system. In 1980, however, before the relaxation of the cross-ownership rule, Glenn Bergland and the Lake Mills Telephone Company were able to get a waiver to the rule which allowed the installation of cable TV in Lake Mills. When the cable company was sold to the RTA, a waiver was again required for a public notice of intent by the phone company, a notice, so to speak, giving first choice to any outside company for the cable TV installation. There was no response to the public notice and Lake Mills Telephone became the proprietor of Lake Mills Cable Television.

In December, 1981, Northwestern Bell offered its toll cable facilities in the Crystal Lake and Woden exchanges for sale to the Cooperative. The Cooperative accepted the offer on the condition that the Cooperative could buy from Northwestern all the long distance toll facilities connected to its exchanges within the Cooperative's operating area; Northwestern accepted the condition, and over the next two years, the purchase was made.

Briefs - 1981

RTA technicians Norman Risius and Alan Swenson returned in February from Albuquerque, New Mexico, where they took a two-week electronics course in the installation and maintenance of Lenkurt digital and analog toll carrier systems, five of which have been purchased by the Cooperative.

—

By March, twenty-two and one-half percent of Cooperative phones were touch tone.

—

The RTA received a thank-you note from a former customer who moved to Missouri and found "my phone service is inferior to the RTA's."

—

The Continental Telephone Company, Iowa's third largest telephone company, received approval from the Iowa Commerce Commission to rent only black telephones to its new customers, as opposed to offering colored phones as previously.

—

RTA employees John Hanna, Neal Sletten, Todd Ranes, Foster Hartman, Wayne Nyguard and Marlin Senne received certificates of achievement for successfully completing electronic and computer evening courses at NIACC.

1982

On January 4, the Cooperative employed Terry Wegener to be its Office Accountant and supervisor of the Cooperative's bookkeeping office staff, a position he holds today in 1990. Terry is in charge of billings and receipts; monthly, quarterly and annual financial reports, and other related bookkeeping functions. Following his graduation from high school in Graettinger, Iowa, Terry attended the American Institute of Business in Des Moines; and for four years was an accountant/auditor with a CPA firm in Des Moines, Iowa. Terry and his wife, Maryann, are the parents of a daughter and a son.

On January 8, 1982, the government ended its long anti-trust suit against the American Telephone and Telegraph company. AT&T agreed to divest itself of its twenty-two Bell operating companies, one of which was Northwestern Bell Telephone serving this area. It kept its long distance lines, its Bell Laboratories, and its equipment manufacturing subsidiary, Western Electric Company. For the story of the government's years-long law suit against AT&T, see Appendix C on page 227.

On February 11, 1982, in an appearance before the U.S. House of Representatives Subcommittee on Telecommunications, Consumer Protection and Finance, Manager Lein, representing OPASTCO, gave testimony relating to the AT&T settlement.

Foreseeing a diminished share of toll pool revenues for the Cooperative in the face of new long distance competition to AT&T, Manager Lein told the legislators that "rural telephone companies must be permitted to expand their revenue potential through enlightened legislation and regulatory policies that permit us to benefit more and more from advanced technology; such as, digital switching, satellite and video communication. Historically, rural consumers have been underserved from the standpoints of both entertainment and non-entertainment telecommunications."

The Cooperative's 1982 annual meeting was September 22 in Lake Mills. In his remarks to the members, Manager Lein said that technology was bringing new telecommunications equipment on the market that greatly increases the scope of communications and the ease of using it. "We have automatic dialers that memorize and redial numbers for you with the push of a button. We are selling cordless telephones and shortly, you will direct dial your credit card calls without the assistance of an operator," Manger Lein said.

"Long distance competition is expanding significantly in urban areas, but these new long distance carriers are not paying their fair share of their use of the local exchanges, which lets them price under the AT&T toll rates.

"Simply stated, competition in the urban areas and intercity long distance will mean higher rates for local telephone service. How can that be, you may ask. Isn't competition supposed to drive prices down, not up?

"Telephone companies traditionally have used a portion of long distance revenues to help support the cost of local service, so that your charge for local telephone service has been underpriced by twenty-five to thirty-five percent from actual cost. Until just recently, the large telephone companies operated as monopolies under public regulation in the public interest. But such telephone service is no longer a monopoly, and competition in long distance is just the start of competition in almost all facets of the telephone business. The large 'Fortune 500' businesses and others using high volume, low-cost intercity toll routes will be the chief beneficiaries of competition; Winnebago's rural area will not benefit.

"Competition will force your Cooperative, as well as other telephone companies, to cost-based pricing which will eliminate the contribution which is made from long distance revenues to local service costs.

"Local companies such as Winnebago cannot go out and offer *only* long distance service *only* in the most profitable areas. Small cooperative telephone companies must work toward the goal of providing universal service. Federal government studies show that present trends in federal competition and deregulation policies, coupled with inflation, will result in local telephone rates going up fifty to seventy-five percent over the next five years," Manager Lein said.

As reported by Manager Lein, Cooperative revenues for 1981 were $1,473,104.75, with expenses of $1,155,416.83, including $292,934.34 reserved for depreciation and amortization, leaving $317,617.92 for debt repayments, dividends and surplus. The REA debt stood at $1,902,615.38.

The members re-elected incumbent board members Keith Finer, Milford Helgeson and Lloyd Kronemann to three-year terms. The board re-elected Vernon Jacobson as President, Victor Knudtson, Vice President and Lloyd Kronemann, Secretary-Treasurer.

Cooperative Purchases Lake Mills Telephone Company

Announcement was made on October 27, 1982, of the purchase of the Lake Mills Telephone and Cablevision Companies by Winnebago RTA from Glenn Bergland. Mr. Bergland had owned the Lake Mills Telephone Company since 1961. In 1980, he made cablevision available to the people of Lake Mills and incorporated it into a company managed by, and part of, the telephone company.

The Lake Mills Telephone Company increased the number of RTA phones by 2,125 to approximately 11,200.

All Lake Mills Telephone personnel became employees of the Cooperative. These employees were:

Marlo Durby, Local Manager of the Lake Mills Telephone Company, who became the Cooperative's Staff Assistant; Paul Sorenson, Installer and Repair Technician of Lake Mills Telco, who became the Cooperative's dial switching technician; Sue Rikala, Lake Mills Telco Cashier-Receptionist/Billing clerk, now in the same position at the Cooperative; and Jan Hendrickson, Lake Mills Telco Clerk/Secretary, who stayed in the same position with the Cooperative only until the two business offices merged.

Employees of the Lake Mills Telephone Company at the time it was purchased by the Cooperative: Paul Sorenson, Sue Rikala, Janice Hendrickson, Marlo Durby. (1982)

"Besides the benefits of increased size and capital base of the Cooperative, it now will have full control over the cost and type of switching service heretofore provided our rural customers by the Lake Mills exchange," Manager Lein said.

In a letter "to the people of the Lake Mills Community," printed in the *Lake Mills Graphic* newspaper, Manager Bergland gave a " 'special thank you' for the twenty-one years of fine cooperation you have extended to us; and we are especially thankful to our excellent employees who have expertly and diligently made our service possible. . . . We are proud of our modern electronic service and rates. . . . True enough, it was a tough decision; but I had to consider my age and that our son is well entrenched in research at Bell Telephone Laboratories and it will not be economical for him to return, even if his heart is here in Northern Iowa. . . We have reasoned that this is the best arrangement to continue good service to this fine community." The letter was signed, "Glenn Bergland and Family."

* * *

Thus, Glenn Bergland brought to a close forty-three years of dedicated service to the people of Winnebago County and adjacent areas, starting in 1939, when as a young man just out of college, he was the driving force in the birth and growth of the Winnebago Rural Electric Cooperative, "lighting up" the countryside for miles around, and continuing through 1949-1973, during which time he again was the driving force in the birth and growth of the Winnebago Rural Telephone Cooperative, bringing modern dial telephone service to rural areas whose phone service, if any, was 'whoop-and-holler' over multi-party, 'listen-in' run-down lines. And then, too, from 1961, when he purchased the run-down Lake Mills Telephone Company and modernized it to an exemplary state-of-the-art telephone system, adding, along the way, cable television for the enjoyment of the citizens of Lake Mills.

* * *

An interesting sidelight on the Cooperative's purchase of the Lake Mills Telephone Company relates to the cross-ownership waiver required for the Cooperative's purchase of the associated cablevision company.

As explained earlier, when the Lake Mills Telephone Company installed cablevision, a cross-ownership waiver was obtained. This cross-ownership waiver, however, was not transferable to the Cooperative, which had to file with the FCC for a new waiver.

Such filing required a thirty-day public notice of the Cooperative's intent to buy Lake Mills Cablevision. It was a genuine fear of the Cooperative that an outsider company would also file with the FCC during the thirty days, saying, "We have the ability to serve Lake Mills with cable television, and we don't think the cablevision should be sold to a telephone company."

In the event, no outsider filed and the Cooperative obtained the cross-ownership waiver.

Also interesting regarding the purchase of the telephone company: To be consistent with the Cooperative's policy that each subscriber pay a ten dollar membership fee, the Lake Mills customers each were asked to pay the ten dollar fee—which they did, one hundred percent.

Bergland High School Fund

In December, 1982, Mr. and Mrs. Glenn Bergland established a perpetual scholarship fund for graduates of the Lake Mills High School. The Berglands donated 2,000 shares of Winnebago Industries stock to fund the program. The stock value was thirteen dollars a share. The high school immediately converted the stock to $26,000 in cash.

Lake Mills High School graduates who plan to continue their education in the vocational, academic natural science field, such as the sciences and biology, are eligible for consideration as recipients of funds from the program.

Briefs - 1982

Beginning February 10, the Cooperative will print its own customer toll charge statements, a service that Northwestern Bell had discontinued doing.

—

A rifle or pistol shot completely destroyed the electronic carrier repeater equipment in its metal housing two and one-half miles east of Kensett on March 28th, and cut off long distance service to Grafton residents. The Cooperative's repair cost was $600.00.

—

Glenn Bergland, Owner/Manager of the Lake Mills Telephone Company, in February gave documented proof to the Lake Mills City Council that the Lake Mills Telephone Company had been in continual existence since before October 1, 1897, and hence has a Lake Mills franchise in perpetuity and is not subject to the twenty-five year renewal provision of the Iowa franchise law. This law exempts utilities established prior to October 1, 1897.

—

Recently, the FCC authorized several United States satellite companies to beam television programs directly from satellites to owners' homes.

Effective this March, RTA customers can direct dial, without charge, the RTA business office, 1-515-592-6105, for business purposes or to report phone trouble when the call is placed from an RTA phone. Formerly, when calls were made to the business office from an RTA exchange other than Lake MIlls, there was a toll charge or the subscriber could call collect.

In June, the Cooperative began entering subscribers' telephone numbers, calling card numbers, and personal identification numbers into the U.S. Intelco data base. Thus, subscribers can direct dial their own calling card numbers from any phone in the nation without operator assistance and resulting surcharge.

This 1982 photo shows Terry Wegener, RTA Office Supervisor, center, with independent auditors Zane Juncker, left, and Bill Christy, of Austin, Minnesota. The Cooperative's accounts are audited annually by direction of the lending agency of the REA.

1983

Deregulation of the telephone industry by the FCC following the settlement of the government's anti-trust suit against AT&T directly affected the Winnebago RTA in a number of ways, including:

(a) A direct charge is made to the customer by the Cooperative for telephone sets and wiring as opposed to formerly including this cost as a part of local service dialtone.

(b) By FCC ruling effective May 25, 1984, multi-line business customers were assessed a monthly "FCC Access Line Charge" of $6.00; and effective June 1, 1985, single-line business and residential customers were assessed a monthly FCC access line charge of $1.00. Today, 1990, the charges are $6.00 and $3.50, respectively. These charges were used to reduce the interstate toll rates charged Cooperative subscribers.

(c) The AT&T "pool" of long distance revenues was discontinued. From it the Cooperative had received payments by a formula based upon revenues and messages.

(d) Some of the above revenues received by the Cooperative were replaced by (1) settlements from the National Exchange Carrier Association (NECA) for interstate toll traffic, which received its revenues from the FCC access line charges. These are Carrier Access Billings (CABS) charges made by the Cooperative to the interexchange carriers—AT&T, U.S. West, Telecom USA, U.S. Sprint, MCI and Iowa Network Services—for interstate usage and remitted to NECA, and by (2) Cooperative billing of CABS charges to interexchange carriers for intrastate usage, which revenues are kept by the Cooperative.

(e) CABS billings are for interstate and intrastate access charges, based on minutes of use and the distance of the Cooperative's transport facilities.

"In*tra*LATA" calls are calls made by the Cooperative's Iowa subscribers within the 515 area in Iowa and calls made by the Cooperative's Minnesota subscribers within the 507 area in Minnesota. Examples of in*ter*LATA calls are those from the 515 area to the 712 and 319 areas in Iowa, and from the 507 area to the 612 and 218 areas in Minnesota. (The "Local Access" charge on the Cooperative's customers' bills is for the cost of the circuit between customer's premise and dial switch office, and switching customer's phone calls to local and extended area service points.)

(f) The purchase price of telephone sets was deregulated.

(g) Effective 1983, calls made from a public phone no longer could be billed to a third-party number unless operator verified the charge with the third party.

(h) The FCC ruled that customers' phone bills be itemized in detail including charges for terminal equipment, wiring maintenance, local access, and FCC access line charges.

(i) Customers can choose between paying a monthly charge of fifty cents for inside wiring maintenance or paying a minimum $40.00 service charge for any maintenance repair required.

Iowa Network Services

In the fall of 1983, the Iowa Network Service (INS) Corporation was founded by the Cooperative and a group of about eighty Iowa independent telephone companies.

Its purpose was twofold—first, to provide centralized equal access service so that competitive long distance carriers would find it attractive to serve the rural areas and small towns that, heretofore, they had chosen not to serve because of their size and remoteness; second, INS would be a long distance carrier, providing both in*ter*LATA and in*tra*LATA services.

Manager Lein was elected Vice-President of the newly formed INS three-person board of directors. The Cooperative paid a membership fee of $2,500 to INS whose profits will be distributed to members. To date, 1990, 128 Iowa telephone companies have become stockholders of INS.

* * *

At its annual meeting in February, Manager Lein was elected president of OPASTCO, the Organization for the Protection and Advancement of Small Telephone Companies, whose members are 400 small telephone cooperative and commercial phone companies with 1,100 exchanges from coast to coast.

(At OPASTCO's June meeting, Manager Lein stated that costly regulation of small telephone companies should be streamlined. "My home state of Iowa became a leader in deregulation when, in 1963, the Code of Iowa was amended to exempt from rate regulation by the Iowa Commerce Commission mutual, cooperative and independents with fewer that 2,000 subscribers, and in later years this was amended to include systems with fewer than 15,000.)

At the cooperative's 1983 Annual Meeting on September 21, Manager Lein reported 1982 revenues of $1,706,914.32 and expenses of

$1,242,363.57, including $302,324.52 reserved for depreciation and amortization, leaving net earning of $464,550.75 for loan payments, dividends and surplus. Long term debt stood at $3,322,741.76.

Incumbent board members, Vernon Jacobson and DeVere Steffensen were re-elected to three-year terms by the membership. The board re-elected Vernon Jacobson as President, Victor Knudtson, Vice President and Lloyd Kronemann, Secretary-Treasurer.

At the meeting, the Cooperative's Articles and By-laws were amended to include cablevision and other telecommunications to the services the Cooperative can offer. "We're in the telecommunications business, a much broader type of business than simply telephone," Manager Lein said.

In October, 1983, Manager Lein spoke before members of the United States Independent Telephone Association, an association of large and small telephone systems, the name "Independent" designating them as separate from the Bell Companies. Manager Lein's remarks are of interest because they speak to the similarity of goals of both large and small telephone companies in the confusing regulatory environment that prevailed following the AT&T break up. They are reprinted as Appendix A of the book, page 223.

With the installation by Northwestern Bell of fiber optic cable from Mason City through the Manly, Kensett and Northwood area, initiated in November, the Cooperative's board of directors—always at the forefront of technology—decided that the Cooperative should install fiber optic cable from Kensett to Joice to Lake Mills. Fiber optics transmit messages converted in laser pulses over thin glass fibers. A single fiber only a sixteenth of an inch in diameter can carry the same number of signals as a copper cable of three to four inch diameter, containing over 1,000 pairs of copper wires. See Appendix D, page 233, "Fiber Optics."

Briefs -1983

The federal excise tax on customers' phone charges was increased to three percent from one percent on January 1, 1983.

—

The Conger, Twin Lakes and Emmons exchange subscribers will pay a six percent Minnesota sales tax, raised from five percent, effective January 1, 1983.

—

RTA employees, Marlo Durby and Paul Sorenson, are conducting evening classes for the RTA's technicians in the basics of the Lake Mills cablevision system. Subjects include construction, coaxial cable connections and maintenance.

—

RTA crews completed the changeover of single-line residence and business telephones to plug-and-jack type, as requested by the FCC. The plugs allow RTA customers to quickly install or change their own phones.

—

The color of the uniforms of the RTA technicians has been changed from green to blue, the standard independent telephone color.

—

RTA technicians Norman Risius and Daryl Stelter are pictured in the April *Dial Tone* replacing the lead-calcium batteries in the Grafton dial office. The batteries filter out circuit noises and serve as standby power source in case of a commercial power failure.

—

May 16, 1983, was the employment date for Regina Adams, rural Lake Mills, to the billing department.

—

Following the Cooperative's Christmas dinner on December 4, 1983, Marlo Durby and Marlin Senne received twenty-five-year service pins from Manager Lein.

—

The cover of this year's Directory features a photo of horses in a grassy woodland grove, courtesy of the photographer, John Carlson, Lake Mills, Iowa.

1984

The Cooperative finalized new long distance contracts with U.S West (of which Northwestern Bell Telephone is a subsidiary) and AT&T. "Hundreds of hours of our time have gone into negotiating new provisions caused by access charges and the divestiture of AT&T," Manager Lein said.

In February, the Cooperative's board of directors gave the go-ahead to management and staff to begin a long range plan that would include new digital switching, cellular telephone systems, consolidation of long distance service and toll ticketing into one centralized switch in Lake Mills, and participation in what came to be the Iowa Network Service (as described in the history of 1983).

In April, Manager Lein was elected to the board of directors of the United States Telephone Association, which has a membership of 1,100 small and large telephone companies. USTA maintains a staff of thirty-five in Washington, D.C., and plays a major role in influencing the FCC and lobbying Congress.

In August, Staff Assistant Durby attended a meeting of National Exchange Carrier Association members in Whippany, New Jersey, to discuss pooling, reporting and cash flow problems that beset average schedule companies like the Cooperative.

Cable Television for Thompson

Given the experience of the Cooperative in cablevision and the results of feasibility studies, the Cooperative made a decision to seek a franchise to install cable TV in Thompson. In a November, 1984, election, Thompson voters granted a cablevision franchise to the Cooperative by a vote of 616 to 26, and in November, 1986, the Cooperative commenced cable TV installation in Thompson.

In an election on November 27, 1984, Grafton and Fertile residents, by a vote of 56-0 and 27-1, respectively, renewed the Cooperative's franchises in the two towns.

The Cooperative's 1984 Annual Meeting was held September 19. In his remarks to the members, Manager Lein said that the Cooperative and 1,400 other independent telephone companies, large and small, serve sixty percent of the nation's phone users and the Bell companies serve the rest.

Manager Lein also cited the threat of "by-pass", where large corporate phone users set up their own communications systems, using microwave, cable, and satellite, thus depriving the regular phone system of revenues, which results in the smaller phone companies, including the Cooperative, having less revenue to maintain the system.

Manager Lein reported 1983 revenues of $2,408,337.06, and expenses of $1,634,504.74, including $435,951.78 reserved for depreciation and amortization, leaving $773,832.32 for debt repayment, dividends and surplus. REA debt stood at $2,172,482.40.

Incumbent board members Victor D. Knudtson, Lake Mills, and James Petersburg, Hanlontown, were re-elected to three-year terms. The board of directors re-elected Vernon Jacobson as President, Victor Knudtson, Vice President, and Lloyd Kronemann, Secretary-Treasurer.

Briefs - 1984

In a 1984 speech to the Minnesota Telephone Association, Manager Lein related the following story, originally told by Lark Miller, the colorful executive of the Alabama-Mississippi Independent Telephone Association.

> When the Occupational Safety and Health Administration (OSHA) requested more information from an injured worker, he wrote as follows:
>
> Dear OSHA: I am writing in response to your request for additional information. In block three of the accident form, I put 'poor planning' as the cause of my accident. You said in your letter that I should expand on this and explain more fully. I trust the following details will be sufficient.
>
> I am a bricklayer by trade. On the day of the accident, I was working alone on the roof of a six-story building. When I completed my work, I had about 500 pounds of bricks left over. Rather than carry the bricks down by hand, I decided to lower them in a barrel by using a pulley, which fortunately was attached to the side of the building at the sixth floor.
>
> Securing the rope at ground level, I went up to the roof and loaded the bricks into the barrel that I had swung over the edge of the roof. Then I went back to the ground, untied the rope and held it tightly to insure a slow descent of the 500 pounds of bricks. You will note in block eleven of the accident form that I weigh approximately 155 pounds. Due to my surprise of being jerked off the ground so suddenly, I lost my presence of mind and forgot to let go of the rope. Needless to say, I proceeded at a rather rapid rate up the side of the building. In the vicinity of the third floor, I met the barrel coming down. This explains the fractured skull and the broken collar bone.

Slowed only slightly, I continued by rapid ascent, not stopping until the fingers of my right hand were two knuckles deep into the pulley. That accounts for my broken fingers.

Fortunately, by this time I had regained by presence of mind and was able to hold onto the rope in spite of my pain.

At about the same time, however, the barrel of bricks hit the ground and the bottom fell out of the barrel. Devoid of the weight of the bricks, the barrel now weighed approximately fifty pounds. I refer you again to my weight in block eleven. As you might imagine, I began a rapid descent down the side of the building. In the vicinity of the third floor, I met the barrel coming up. This accounts for the two fractured ankles and the lacerations of my legs and lower body.

The encounter with the barrel slowed me enough that my injuries were lessened when I fell onto the pile of bricks and only three vertebrae were fractured.

I am sorry to report, however, that as I lay there on the pile of bricks in pain, unable to stand and looking up at that barrel six stories above me, I again lost my presence of mind—and let go of the rope. . .

—

An addition to the Lake Mills warehouse is under construction.

—

The Cooperative has purchased a Texas Instruments business computer and will now do its own billing functions instead of contracting them out.

—

The Cooperative paid $75,183.32 in county property taxes for tax year 1982-1983, with Winnebago County getting the largest tax, $38,375.52, and Mitchell County the smallest, $608.61.

—

The offices of the former Lake Mills Telephone Company on north Washington were moved to the Cooperative's general offices on the east side of Lake Mills in April. The RTA will retain the south portion of the Lake Mills Teleco building which houses dial switching equipment. The balance of the building was leased to Pioneer Savings and Loan for use as office space.

—

Foster Hartman assumed the title of System Engineer for the RTA on July 1, 1984.

—

The October *Dial Tone* reports on "Hypnotism By Phone":

"Dial-a-Trance, a new dial-it service available in the Chicago area, offers a five-minute recording that puts callers in a hypnotic state. The hypnotist, Larry McManus, says 50,000 people have called the number since January. 'Dial-a-Trance makes people feel good about themselves and helps them to relax,' says McManus. On the tape, his voice tells callers to sit or lie in a comfortable position. After counting from eight to one and giving instructions for concentrating on breathing, he gives suggestions he says make people feel 'refreshed, relaxed and wonderful' when they awaken."

—

The cover of this year's Directory features the photo, "Giant Canada Goose Family," by John Carlson, Lake Mills, Iowa.

—

Cellular Telephone and the Cooperative

In June, 1984, the Cooperative, the Clear Lake Telephone Company and the Farmers Mutual Telephone Company, Nora Springs, formed a new corporation to provide cellular radio telephone service to the Cooperative's current service area; and also to Albert Lea, Austin, Manly, Northwood, Mason City, and other nearby communities.

Cellular radio telephone differs from conventional mobile radio telephone systems in numerous ways.

First, only twenty-one channels are available in the mobile radio system; and when these are used in one area, they cannot be reused in a nearby area.

Cellular systems have 333 channels available in each cell. A cell is a transmitting station, which can be reused in nearby areas. Each call is automatically assigned until all channels are busy.

In the mobile system, only a few subscribers can use the system at a given time, and once the system is fully loaded, further growth is impossible. In contrast, a cellular system maintains wireline communication links to many cell sites. Each site is a base station, similar to the central station in a conventional mobile system; but each cell site covers a smaller area. As a cellular telephone user moves about, or roams, the system keeps track of the user, switching the call, uninterrupted, from cell to cell as the roamer moves through the respective areas. Further, each channel is reused to make maximum use of its frequency allocation.

Cellular licenses are granted by the FCC. The FCC, in 1988, divided the United States into several hundred metropolitan areas and several hundred rural service areas (RSAs) within which cellular service is provided by the area's licensees.

An RSA consists of a five to six county area with a population of 100,000 to 200,000. Any wire line telephone company can file for a cellular license in an RSA where it has a presence. The Cooperative has a presence in RSA 10 in Minnesota and RSAs 13 and 14 in Iowa, and filed for licenses in each. The FCC, however, decided to grant licenses to approved applicants by lottery.

To increase its chances for license award and also to share the cost of a cellular system and thus reduce the risk, the Cooperative formed consortiums (an FCC approved procedure) with about sixteen telephone companies in RSA 10 in Southern Minnesota, and with eighteen telephone companies in RSA 14 in Iowa including Winnebago County and Mason City, and with eleven companies in RSA 13 in Iowa, east of Mason City.

The lotteries were held in the summer of 1989, and the Cooperative was the winner in RSA 14, but not in RSAs 10 and 13. Because of the consortiums, however, the Cooperative will be a limited partner in RSAs 10 and 13, and general managing partner in RSA 14, and as such, together with Clear Lake Telephone and Farmers Mutual, Nora Springs, it will have more control both financially and over operations in RSA 14 than it will have as a limited partner in RSAs 10 and 13.

Again, though, with its seeming predilection for the laws of chance, the FCC has taken another band of frequencies and will allocate it, again by lottery, to "non wire line carriers", which include anyone *but* a telephone company, be he a farmer, businessman, clerk or entrepreneur of any kind. This means that phone companies will not be the only cellular entities in the RSAs; there will be others providing competitive cellular service.

Currently, 1990, "We're putting together financing for the three systems, determining cell site locations, completing design and engineering specifications," Manager Lein said.

On August 15, Hans Olsen, Lake Mills, was employed as buildings and grounds maintenance and truck cleanup.

1985

The Iowa State Commerce Commission did a survey of Iowa's 167 independent telephone companies to ascertain degrees of customer satisfaction. The RTA rated high in the survey; ninety-eight percent of its customers rated service excellent or good.

The Cooperative received approval from the REA for a loan of $7,522,200 at nine percent interest to replace all fifteen of its dial switching offices with digital electronic equipment. To date, 1990, however, the Cooperative has been able to generate sufficient general funds and savings to proceed with the conversion without having to draw against its loan credit.

Explosion at Hanlontown

On Monday, April 8, 1985, the RTA's Hanlontown dial office was blown to smithereens by a natural gas explosion. No one was in the building at the time and, fortunately, those in the vicinity were far enough away from the explosion at 4:30 that afternoon to escape injury from the flying debris. The nearby post office, the Corner Inn restaurant, and a retail store had windows broken by the concussion and flying bits of concrete.

The building was electrically heated and did not have gas service, but there was a leak in an underground gas line that was located in the alley near the building, and the escaping gas followed telephone cable conduit into the building where a spark from the switching equipment ignited the gas.

In less than an hour, Cooperative crews were on the scene. They furnished makeshift telephone service to Hanlontown's 213 residents by running four emergency lines to the town's community center. Callers to Hanlontown were answered with a recording that instructed them to call any of four emergency numbers, which were answered by personnel in the center.

Meanwhile, other RTA crew cleaned up and hauled away the wreckage.

By noon the following day, Tuesday, the Cooperative had located a temporary digital dial switch, twenty-four replacement batteries, a charger and cable distribution frame, and numerous additional pieces of equipment to be shipped from Iowa, Illinois, Texas, and South Dakota.

Co-op crews next extended a cable to an 8 x 10 foot chicken coop-shaped shed where they installed the equipment. At 5:30 p.m. on Wednesday, a portable, trailer-mounted digital switching service arrived from the Golden West Telephone Cooperative in Wall, South Dakota, in quick response to the Cooperative's call for help.

Within fifty-six hours of the explosion, the first customers were called on the temporary switch, and by 2:00 a.m. Thursday, all customers had dial tone and could place local and long distance calls, and extended area service calls to Fertile. The last electronic equipment arrived from Texas on Friday morning, and EAS trunks to Manly and Joice were turned on.

Before being blown to smithereens: Hanlontown switch building.

The Cooperative took bids on new switching equipment, with Northern Telecom, Raleigh, North

The wreck from a heck of an explosion. Gas leaking from an underground pipe beneath the RTA's Hanlontown switch building followed a telephone cable into the building, where it was ignited by a spark. (1985)

Carolina, the low bidder. Installation was completed on May 15, and the portable switch trailer was returned to Golden West.

The non-stop work of the Cooperative's dedicated workers in restoring service to Hanlontown in record time was recognized both by many letters from grateful Hanlontown residents and by write-ups in telephone trade journals.

* * *

On April 29, 1985, Victor Knudtson resigned as a director of the Cooperative, effective May 1. Mr. Knudtson had served on the board for more than eighteen years, both as secretary-treasurer and as vice president, a position currently held. In a notice the members he said, "I have always considered it a great privilege to be associated with such a fine organization. I wish the very best to my successor and to the Corporation itself."

Manager Lein said, "Victor has given a great deal of time to guide and direct the Cooperative through the many complex changes. Directors, employees and members owe Victor a big thanks for his work."

The board of directors appointed Jerry Epland, rural Emmons, to succeed Mr. Knudtson on the board. At the same time, they elected Milford Helgeson as Vice President of the board.

At its July board meeting, the Cooperative's board of directors accepted the resignation of Dudley Weible as the Corporation's attorney. Mr. Weible had served the Cooperative since its beginning in 1950, and the Winnebago Rural Electric Cooperative since its beginning in 1939. "Dudley Weible has been very instrumental in the growth and success of the RTA and the Electric, and we shall greatly miss Dudley and his valuable counsel," Manager Lein said.

The Cooperative's board subsequently employed Mr. Weible's law firm of Bakke, Petersen and Peterson to represent the Cooperative through attorney Dennis Peterson. (The law firm today, 1990, is Petersen & Peterson.)

The cooperative's attorney, Dennis Peterson, of the law firm of Petersen & Peterson, Forest City

The Cooperative's 1985 Annual Meeting was held September 18. In his remarks to the members, Manager Lein said the telephone industry has "experienced an explosion with the breakup of AT&T on January 1, 1984, somewhat akin to the Hanlontown explosion . . . The short range effects of divestiture, competition and deregulation have been public confusion and concern. And no wonder! What can we expect when we ask an entire nation to please bear with us while the FCC goes about restructuring telephone service that has suited Americans for 109 years! Your Cooperative has embraced the new technologies and will use them to increase the production of our networks and our employees."

Manager Lein reported 1984 revenues of $3,101,438.03, and expenses of $1,858,556.69, including $585,703.38 reserved for amortization and depreciation, leaving $1,242,881.34 for debt repayment, dividends and surplus.

RTA members re-elected incumbent directors Keith Finer, Milford Helgeson, and Lloyd Kronemann to three-year terms, and Jerry Epland to a two-year term. The directors re-elected Vernon Jacobson, President; Milford Helgeson, Vice President; and Lloyd Kronemann, Secretary-Treasurer.

Testifying before the U.S. House of Representatives subcommittee on regulatory issues: Manager Lein, RTA, and Lorenda Ackley, Taconic Telephone Corp., Chatham, New York.

On October 16, 1984, Manager Lein was re-elected to the board of directors of the United States Telephone Association.

RTA crews completed the installation of 2,000 feet of fiber optic cable in the town of Kensett in October. The cable, which carries up to 8,000 phone calls simultaneously, joins Northwestern Bell's Mason City-Kensett fiber cable link.

$243,463.99 to Members

On December 1, 1985, the Cooperative distributed $243,463.99 to members which was the payment of deferred dividends for the years 1962 through 1967.

Briefs - 1985

"Your crews acted in a professional manner and showed a sincere concern for the needs and safety of our community," read one of the many letters of thanks to the Cooperative after the explosion at Hanlontown.

—

Suicide Helpline of Iowa is an organization in Clear Lake that operates a twenty-four hour toll-free, state-wide suicide help line. In 1984, there were 300 reported suicides in the state and over 5,000 attempts. (Note: As of 1990, there is no longer a Suicide Helpline telephone number. The Crisis Intervention Center at Mason City, Iowa, has a twenty-four hour number that can be called collect, 515-424-9133, and there is a Rochester, Minnesota, suicide help line, 1-800-638-HELP.)

—

In 1985, there were 40,000 cooperatives, customer owned, in the fields of electric, telephone, housing, food, insurance, farm credit, farm marketing and farm supply, health and day care, and credit unions.

—

The RTA continues to rent its building just west of its former Lake Mills business office to GTC Auto Parts.

—

Kruger Construction, Northwood, was the low bidder for the Cooperative's new dial office building in Hanlontown.

—

Kensett residents voted 134 to 5 to renew the Cooperative's Kensett franchise on November 5, 1985.

—

Foster Hartman, RTA System Engineer, and Norman Risius, RTA Dial Technician, received twenty-five year service awards and congratulations from Manager Lein at the Cooperative's December Christmas Dinner. Both men started with the Cooperative in 1960 and helped with the transition from multi-party to all-private lines, and from aerial poles and wire to all-underground cable.

—

The cover of this year's Directory features the photo of a Monarch butterfly, entitled, "The Last Stop," by John Carlson, Lake Mills, Iowa.

1986

One of two of the Cooperative's cable TV receiving/sending systems, this one at Thompson. There are three dishes (third dish out of picture at right), a tower and a control building. The tower receives off-the-air signals, ABC, NBC, CBS and IPBN, from Rochester, Austin, Mason City and Des Moines, respectively. The dishes capture satellite signals for all other chanels, and the control building consolidates the signals and sends them out on cable to participating TV sets.

Construction of the Thompson cable television system began in July with the erection of a 120-foot tower and the completion of the cement block building for the electronic equipment. Cable laying and testing were completed in October, and by November all Thompson cable customers were receiving cable service.

The Cooperative's 1986 Annual Meeting was September 17. In his remarks to the members, Manager Lein said that "today small companies like ours see dangers in the universal rush to alter the cost recovery process by shifting too much of the cost burden from long distance carriers to local exchange customers. We also fear that competition in urban toll service could threaten uniform, nationwide long distance pricing. Along with the other industry leaders, I filed petitions with the FCC asking them to reduce tariff filings and regulatory burdens on small telephone companies because we 'make up an insignificant fraction of the exchange carrier industry and have virtually no ability to exercise market force in the interstate market.' . . . The 1,100 smallest telephone companies with fewer than 20,000 access lines make up only two and one-half percent of the market."

Manager Lein reported 1985 revenues of $2,901,435.03 and expenses of $1,896,079.54, including $548,832.74 reserved for amortization and depreciation, leaving $1,005,355.49 for debt repayment, dividends and surplus. As of December 31, 1985, dividend credits and surplus totalled $5,975,423.38 and long term debt stood at $1,640,639.25. The Cooperative had 11,200 phone subscribers and 650 cable TV subscribers.

Incumbent board members DeVere Steffensen and Vernon Jacobson were re-elected to three-year terms; and the board re-elected Vernon Jacobson as President; Milford Helgeson, Vice President; and Lloyd Kronemann, Secretary-Treasurer.

* * *

In May, 1986, the Cooperative accepted a bid of $1,505,113 from Northern Telecom for new digital switches and installation in the Fertile, Grafton, Joice, Kensett, and Lake Mills exchanges.

In December, the second digital switch in the Cooperative's multi-million dollar digital conversion was installed in the Lake Mills exchange.

The Cooperative's Lake Mills digital switchboard. (1987)

It will be followed with thirteen more throughout the RTA system in 1987 and 1988. The first was installed in Hanlontown, replacing the switch lost in the explosion.

The Lake Mills installation is computer processor controlled and consists of thousands of electronic cards and chips; it will process all RTA toll ticketing, recording, rating and billing. It also provides for the features of call waiting, call forwarding, three-way conference calling and speed calling.

The twenty-one mile underground fiber optic cable from the Cooperative's Lake Mills office to Kensett is spliced in six locations. The contract splicers used a microscope to join the hair-thin fibers. At Kensett the cable is connected to Northwestern Bell's fiber cable from Mason City. The cable has four fibers, two active and two spare. Each fiber can transmit up to 8,000 voice channels. At the switching center, the fiber's laser light signals are converted into electric signals.

Briefs - 1986

The Rev. Robert W. Bresemann, Hanlontown, was issued the one millionth telephone calling card from the U.S. Intelco Network. The RTA is a stockholder in the USIN which manages the computerized data base that monitors calling cards for its 300 independent telcos nationwide.

—

In April, AT&T auditors reviewed the Cooperative's toll separation records from 1964 to date with Staff Assistant Durby and Office Supervisor Wegener.

—

RTA Technicians, Al Swenson, Paul Sorenson, Norman Risius, Marlin Senne, and Daryl Stelter are enrolled in a five-week manufacturer's course in digital switching at Des Moines Area Community College in Ankeny. They will be responsible for operation and maintenance of the Cooperative's new digital equipment at Hanlontown, and for the fifteen switches scheduled for the RTA system in the next two and one-half years.

—

In April, the Future Farmers of America awarded the Cooperative a citation certificate for helping FFA accomplish outstanding achievements in farming, leadership, citizenship and community welfare.

—

The cover of this year's Directory features the photo, "River of Beauty," by John Carlson, Lake Mills, Iowa.

1987

At 11:00 p.m. Thursday, January 22, the RTA's Lake Mills exchange was cut over to its new digital switch. The RTA's Lake Mills customers awakened Friday to the use of new customer call features that included call waiting, call forwarding, three-way calling, single digit speed dialing and revert call.

On February 12, the Kensett and Grafton exchanges were cut over to digital switching with the same added features.

In January, Manager Lein was elected to the nine-member board of directors of U.S. Intelco. USIN manages a nationwide data base for independent telephone company calling card services and sells computer products, in addition to information services to help streamline the business operations of telephone companies. More than 300 independent telephone companies have signed agreements with USIN to provide billing and collecting services to AT&T and, thereby retain the revenue flow resulting from these services.

The Joice and Fertile exchanges were cut over to digital switch on March 11 and March 25, respectively.

Descrambling Satellite TV

The Cooperative recently became a member of the National Rural Telecommunications Cooperative (NRTC), an organization of electric and telephone systems to provide entertainment and business television services to rural customers.

The Cooperative serves 850 customers with cable TV service, but this is only possible in areas where the subscriber density is high enough to make the service economically feasible. Instead of putting up new satellites, the NRTC is organized to provide descrambled satellite programming for resale to rural areas and small communities beyond the reach of cable systems.

The plan embodies some of the same pioneering spirit and in many ways is reminiscent of the early days of rural telephony.

Today, 1990, the Cooperative has sixty customers buying the satellite program from the Cooperative.

For the rural customers (not living in Lake Mills or Thompson) to participate in the Cooperative's program offering descrambled satellite TV, they must (a) have a satellite dish and receiver and be responsible for their installation and maintenance, (b) have a Video Cipher II descrambler, which can be purchased through RTA, and (c) agree to pay a monthly fee for their choice of satellite programs offered by the RTA. Some programs, of course, are not scrambled and are free.

INS Advantages

In a speech in July, 1987, to the OPASTCO organization in Washington, D.C., Manager Lein explained the principal function of Iowa Network Services. Iowa Network Services stockholders are 136 Iowa telephone companies, of which the Cooperative may be the largest individual stockholder with its investment of about $400,000. "Whether or not our customers enjoy the benefits of interexchange (equal access) competition will no longer depend upon whether or not they are served by a Bell operating company. By connecting 160,000 access lines in 306 communities to a central digital switch, INS makes it possible for competing interexchange carriers to obtain equal access to ninety-five percent of the small independent Iowa Telephone companies through interconnection at the central access tandem in Des Moines. Our customers enjoy the benefits of new offerings, more options, price competition and lower rates."

(Editor's Note. The above is yet another example of the Winnebago RTA's role in telephone leadership for rural area subscribers. The RTA was a founding member of INS, with Manager Lein on its first three-member managing board, and the RTA's installation of state-of-art digital switch equipment and fiber optic transmission enables it to take full advantage of Iowa Network Service.)

On July 15, 1987, funeral services were held for M.J. Madson. Mr. Madson was a founding board member of the Cooperative, serving from 1950 to 1967. After Mr. Madson resigned from the board, he continued to work for the Cooperative in securing rights-of-way and easements.

The Cooperative's 1987 Annual Meeting was held on September 23. In his remarks to the members, Manager Lein expressed apprehension that the federal REA telephone lending program might be discontinued. "This and other current proposals in the Administration's budget would be an absolute disaster for rural America, leaving its institutions and economy in a shambles.

"There is simply no way we can double military spending over five years, cut annual revenues by $135 billion and balance the budget at the same time. These are simply unworkable proposals passed off as credible solutions," Manager Lein observed.

Manager Lein reported 1986 revenues of $3,047,795.95 and expenses of $1,893,369.30, including $539,390.19 reserved for depreciation and amortization, leaving $1,154,426.65 for debt repayment, dividends and surplus. As of December 31, 1986, the Cooperative's long term debt was $1,502,661.75 and its dividend credits and surplus balance was $7,123,722.31.

The members re-elected incumbent Co-op board members James Petersburg and Jerry Epland to three-year terms. The board re-elected Vernon Jacobson as President; Milford Helgeson, Vice President; and Lloyd Kronemann, Secretary-Treasurer.

The Scarville and Leland exchanges were cut over to digital switch on November 10 and November 24, respectively, at 5:30 a.m.

Briefs - 1987

On January 22, Mayor Keith Helland of Lake Mills made the first long distance call over the RTA's new digital switch and fiber optic toll cable.

—

Marlo Durby registers attendants at the Co-op's table, 4-H Fair, August, 1987.

In February, RTA's Marlo Durby and Paul Sorenson attended a Northern Telecom school in Chicago for the operation and maintenance of computerized long distance tape equipment for recording time and billing charges for toll calls. Daryl Stelter attended an NEC electronics class in interfacing fiber cable with digital switching equipment. John Hanna, Todd Ranes, Marlin Senne, Paul Sorenson and Gene Walz recently received training in the emergency repair of fiber optic cable.

—

This summer, Cooperative crews are burying sixty miles of long distance cable in the Woden, Clear Lake, Thompson, Rake, Conger, and Twin Lakes areas. They will modify the dial office buildings in eight exchanges to make room for the installation of new higher capacity emergency generators and air conditioners, the latter for cooling both the rooms and the new digital switches therein.

—

On October 7, the Emmons exchange was cut over to its new computerized digital service. A large picture in the Lake Mills Graphic features Alan Swenson, Norman Risius, and Paul Sorenson at the switchboard. November *Dial Tone* also pictures Daryl Stelter at the switchboard.

—

The cover of this year's Directory features a photo by John Carlson, Lake Mills, of the new computerized digital switch in the Cooperative's Lake Mills switching center.

Temporary fixtures at the Hanlontown explosion site: Converted chicken coop containing cable connections and cable leading to switch trailer on loan from South Dakota.

1988

In March, the FCC approved the Iowa Network Services project, which means that Iowa's rural customers, including the Cooperative's, will have a choice of carriers for their long distance calls, which now are provided solely by AT&T. Said Paul Franzenburg of the Iowa Utilities Board, "This puts the fellow out in the rural farmhouse. . . . .on an equal footing with the man in the Ruan Center, Des Moines."

But long distance carriers complained to the Iowa Utilities Board that INS wanted to charge them "unreasonable" access fees to interconnect with the INS system, and thus create "unconscionable" profit which INS would use to subsidize its own long distance service.

As a result of the complaint, the utilities board reduced the access charge INS can collect on intra-state calls from other phone companies.

On July 19, 1988, the Iowa Utilities Board issued an order prohibiting charges for Extended Area Service (EAS) listings in telephone directories. Traditionally, Iowa and Minnesota telephone companies have exchanged and printed without charge EAS listings in their respective directories. Northwestern Bell, however, tried to levy a forty cents per listing charge in their directories. The independent companies, led by Manager Lein, had contested the Bell levy, through the utilities board, whence came the order.

The Cooperative's 1988 annual meeting was held September 21. In his remarks to the members, Manager Lein said that with its modern digital and fiber optic network, the Cooperative offers advanced customer service to business and residence alike. "But it takes more than networks and technology—it takes employees who believe in tradition, in friendly service, in following through on commitments, and in a desire to please," he said.

Manager Lein reported 1987 revenues of $3,571,979.43 and expenses of $1,992,883.12, including $506,350.01 reserved for depreciation and amortization, leaving $1,579,096.31 for debt repayment, dividends and surplus.

The members re-elected incumbent directors Keith Finer, Milford Helgeson and Lloyd Kronemann to three-year terms. The board re-elected Vernon Jacobson as President; Milford Helgeson, Vice President; and Lloyd Kronemann, Secretary-Treasurer.

In October, the United States Telephone Association re-elected Manager Lein to a second three-year term on USTA's board of directors.

Bergland $250,000 Luther College Scholarship Fund

Children of current, past and future customers of the Winnebago RTA and Winnebago Rural Electric Cooperative who aspire to go to college are eligible to receive a wonderful financial break through the generosity of Mr. and Mrs. Glenn Bergland, Lake Mills, Iowa.

These children (students) are eligible to apply for scholarships of $2,000 or more at Luther College, Decorah, Iowa, if they intend to pursue studies in natural sciences, such as biology, chemistry, engineering, mathematics, physics, premedicine, etc.

The money will come from a scholarship fund of more than $250,000, which will be fully endowed at a later date after the sale of certain trust assets belonging to Mr. Bergland.

What's more, recipients showing exceptional academic progress at Luther College may be eligible for successive grants.

Glenn Bergland organized and managed the Winnebago REC for thirty-five years and the RTA for twenty-five years. Glenn is a 1939 graduate of Luther, and the college scholarship document reads in part, "to commemorate Glenn Bergland's relationship with the college and to express appreciation to all past and present members of the RTA and REC, and in recognition of the value of the natural sciences and a Christian education."

The college will hold the fund's principal in perpetuity, forever insulated from invasion, with the scholarship grants to come from the earnings of the fund's principal. Each year a portion of the earnings will be absorbed into the fund to replace the erosion of inflation. The fund will continually increase over the years.

There currently are more than 6,000 families who are customers of the RTA and REC, and scores who have been, and thousands who will be, into the distant future ad infinitum; and it is their children who are and will be eligible for the Bergland fund scholarships.

* * *

On December 5, a group of Japanese technicians visited the RTA. They represent Japanese who bought from Northern Telecom several million dollars of equipment similar to that purchased by the Cooperative from

Northern Telecom. The technicians will install the equipment back in Japan.

Marlin Senne, RTA plant superintendent, standing, right, with Japanese visitors. (1983)

"The reason the guests were brought here is because we have the largest host satellite switching system provided by Northern Telecom in the U.S. today," Manager Lein said.

"This is the first time to my knowledge the Japanese have purchased millions of dollars of electronic switching equipment manufactured in the United States for delivery in Japan," he added.

(Note: The RTA sets an example not only for American communication companies, but for Japanese as well!)

Briefs - 1988

On January 11, Eileen Schissel began working for the Cooperative. She is the Cooperative's office secretary and also does computer and general office work.

—

Exact time: Dial 1-900-410-TIME (1-800-410-8463) and get the most precise time in the world, given as Universal Time and also as Eastern Standard Time: The hours, minutes, seconds from the U.S. Naval Observatory Master Clock, which is accurate to one ten-millionth of a second. Listeners adjust Eastern to Central time by subtracting one hour, Universal to Central time by subtracting six hours. Universal time is the time at the Prime Meridian in Greenwich, England. The 900 call cost is fifty cents for the first minute and thirty-five cents for each additional minute.

—

The Rake exchange was converted to digital switch February 16, 1988. Rake's toll calls now are routed through Lake Mills and, thence, to Kensett via fiber optic cable where they connect with Bell's Mason City link.

—

Crystal Lake and Woden were converted to digital switch on March 15 and April 12, respectively.

—

The Cooperative's subscribers in its three Minnesota exchanges now have ten cents added to their monthly phone bills for the TACIP, Telecommunications Access for the Communication Impaired Persons, program mandated by the Minnesota Legislature. It provides special telephone equipment for low-income hearing and speech impaired persons.

—

The Cerro Gordo County Board of Supervisors has requested 911 cost information from the seven telephone companies that serve the county.

—

The 911 Minnesota emergency telephone number was activated in the Twin Lakes exchange when it converted to digital switch on May 10. The *Dial Tone* pictures Norman Risius modifying a pay phone in Twin Lakes for 911 calls to be made without requiring a coin.

—

Northwestern Bell Telephone, Montana Bell Telephone and Pacific Northwest Telephone all took the name of U.S. West Communications, effective July 1.

—

Thirty TV channels are now available through the Cooperative's descrambling program for satellite dish owners.

1989

In February, the Cooperative mailed "equal access" ballots to its subscribers, with the exception of five exchanges, for their selection of long distance carriers; one for in*tra*LATA calls (within the 515 area), and one for in*ter*LATA calls (all other long distance calls, calls between *L*ocal *A*ccess *T*ransport *A*reas, whether in the same state or interstate).

The Conger, Twin Lakes, and Emmons exchanges were excluded from the ballot mailing because their calls are routed through the Owatonna exchange, which does not offer equal access. Nor were Forest City rural and Buffalo Center rural sent ballots; their toll calls are routed through General Telephone, which does not offer equal access.

An amended ballot was mailed in March (see below).

The choice of long distance carriers was made possible by two events: (1) the breakup of AT&T and the resulting formation of competitive carriers, and (2) the formation of Iowa Network Services, making it technologically possible to interconnect the competitive carriers with subscribers within the INS service area.

The in*tra*LATA carriers listed on the March amended ballot were Teleconnect (now Telecom USA), MCI Communications, U.S. Sprint, and INS. The in*ter*LATA ballot listed the same carriers, plus AT&T.

The Reason for the Amended Ballot

The AT&T divestiture decree okayed AT&T continued presence in the in*ter*LATA long distance telephone arena; at the same time, it ruled that the Bell Operating Companies, of which AT&T was divested by the decree, could provide in*tra*LATA *but not* in*ter*LATA service. And therein was the genesis of the Cooperative's amended March ballot, which followed the February first ballot.

U.S. West, serving the Cooperative area through what formerly was Northwestern Bell Telephone, declined to have its name on the February ballot listing of in*tra*LATA carriers. To sidestep the U.S. West decision, INS telephone companies put the words "No Change" on the February ballot. Since in*tra*LATA calls by the 158,000 INS subscribers were through U.S. West, subscribers designating "No Change" would in effect cast their ballot for U.S. West.

U.S. West, not wanting to be on the ballot in any case, contested being put on via the side door. It said it would mean paying INS an access fee to use the INS twenty-five million dollar fiber optic system linking rural Iowa with its central switch in Des Moines. In turn, INS argued that "No Change" would protect those end users who wanted to stay with Northwestern Bell, and that U.S. West should not be allowed to abandon more costly rural customers without authorization from the Iowa Utilities Board.

U.S. West did ask the board, which ruled in its favor. Hence, the amended ballot with no reference to U.S. West.

In a speech later that same year on July 24, 1989, to the National Association of Regulatory Commissioners, Manager Lein said that U.S. West should have agreed to be on the ballot. "If the Bell Operating Companies are allowed to pick and choose the area in which they will provide service through the balloting process, then any incumbent carrier could decide to withdraw service from the high cost rural areas by refusing to appear on an equal access ballot," he said.

* * *

RTA employees and directors by law were not and are not allowed to advise customers on choice of carrier.

The Cooperative billings to its customers include AT&T and INS toll charges, and starting in June, 1990, U.S. Sprint. Subscribers receive separate bills from Telecom USA and MCI.

RTA long distance calls are now routed through the Lake Mills switch, thence over the new fiber facilities to the INS equal access switch to Des Moines, where they are transferred to the lines of the pre-selected long distance carriers. The changeover was completed June 14, 1989.

* * *

On March 7, in Austin, Texas, Manager Lein spoke to representatives of the seven regional Bell holding companies who are striving to reduce rate regulation imposed on them by both federal and state regulators. To this end, they have proposed a change from "Rate of Return" regulation to a more flexible plan called "Price Cap" plan.

"Although our Cooperative is not state rate regulated, it is regulated as to quality of service and boundary of service area," Manager Lein told the group.

"This kind of regulation is in itself a form of price cap regulation in that we work to keep our costs low and strive to do better than the average; nor do we spend huge sums on rate cases, which involve accountants, consultants, attorneys, and time of our staff and management."

Manager Lein also said that the U.S. Intelco Network, of which he is a director, is organizing its own SS7 network, Signet 7, for enhanced service, subscriber data bases and application software for its independent telephone company members.

$950,442.91 Paid to Subscribers

In May, the Cooperative mailed 8,931 patronage dividend checks totaling $950,442.91, representing payment of deferred dividends for the years 1968 through 1978.

In August, Wayne Roberts retired from the Cooperative after twenty-two years of service. After the disastrous 1966 ice storm, the Cooperative made the decision to go underground with storm-proof cable replacing its thousand-mile aerial system. Wayne had a substantial background in underground cable construction and he was hired in 1968 as the Cooperative's Construction Superintendent. "Wayne has been successful with all his construction assignments; we greatly appreciate his fine contribution to the improvement of RTA telephone service," Manager Lein said.

The Cooperative' 1989 Annual Meeting was held September 20. In his remarks to the members, Manager Lein said that the Cooperative, because of its Conger, Twin Lakes, and Emmons exchanges, had joined other Minnesota telephone companies to build a centralized equal access system similar to Iowa Network Services; and that authority to construct and operate the system was being requested from both the FCC and the Minnesota regulatory authorities.

Manager Lein reported 1988 revenues of $4,416,678.55 and expenses of $2,179,583.74, including $643,459.75 reserved for depreciation and amortization, leaving $2,237,094.81 for debt repayment, dividends and surplus.

The RTA's revenue came fourteen percent from local area telephone service; seventy percent from charges to network carriers for the use of the Cooperative's facilities for long distance interexchange service; and sixteen percent from providing billing and collecting services to interexchange carriers, sales, lease and maintenance of customer premise equipment, and directory advertising and various rental income.

The members re-elected incumbent board members Vernon Jacobson and DeVere Steffensen to three-year terms. The board re-elected Vernon Jacobson, President; Milford Helgeson, Vice President; and Lloyd Kronemann, Secretary-Treasurer.

In October, Manager Lein was elected to a second term on the executive board of the United States Telephone Association (USTA). USTA's membership ranges from the Bell Atlantic Corporation with nearly seventeen million access lines to small rural telephone companies with as few as fifty access lines.

USTA plays an active role in the formation of federal communication's telephone policy and telecommunication's legislation.

In December, Manager Lein repeated his projection of year-end 1990 for completion of the digital switching conversion for the Buffalo Center rural and the Forest City rural exchanges. (See page 211 for story of the transfer of switching service for the two exchanges to the Cooperative from General Telephone.)

Briefs - 1989

The January *Dial Tone* pictured the RTA volleyball team: Dan Bell, Terry Wegener, Roger Rogstad, Todd Ranes, Gene Walz, Doug Thompson, Daryl Stelter, and John Hanna.

—

Effective April 1, the FCC equal access monthly charge to residence and single line business customers was increased to $3.50.

—

Effective April 1, the Cooperative levied a one-percent tax on telephone charges to its eighty customers living in Cerro Gordo County as mandated by that county's one-percent local option tax.

—

In June, the Cooperative erected two new satellite dishes, one in Lake Mills, the other in Thompson, to receive Midwest Sport Channel signals from the Weststar satellite. The Cooperative's two dishes already in place receive signals from the Galaxy One and Galaxy Three satellites.

—

On June 1, U.S. West reinstated address information to information callers.

—

The thirty-five year history of the National Telephone Cooperative Association, 1954 to 1989, recounted in an attractive printed booklet, featured Glenn Bergland on many pages. Glenn was a founding board member of NTCA and served in the office of vice president and president.

—

The August 14, 1989, *Mason City Globe Gazette* pictured Alan Duea of the Clear Lake Telephone Company testing a cellular phone in his auto, listing cellular phone equipment for a car as costing $900 to $1,500, for an overall average amortized cost of from $80 to $100 per month, plus calling cost of up to thirty-two cents a minute.

—

In September, RTA employee John Hanna was promoted to Construction Foreman. John started with the Cooperative on March 31, 1980.

—

Four Soviet Union citizens , three men and a woman, visited the Cooperative in November. Foster Hartman reviewed the history of the Cooperative with them and explained the Cooperative's modern telephone technology including digital switching, fiber optics and cable television. The Soviets were in the Forest City-Lake Mills area for several days.

—

The cover of this year's Directory depicts the painting, "Mallards at Elk Creek," by artist Floyd Thomas, Lake Mills, Iowa.

Foster Hartman of the RTA describes automatic number identification (ANI) equipment to Mayor Helgren, Thompson, as Manager Bergland, far left, looks on. ANI automatically registers caller's telephone number to the toll charge without operator intercession.

1990

On February 7, Manager Lein testified before the U.S. House of Representatives' Government Operations Subcommittee in a hearing on cable television service. It was the third in a series of hearings to explore issues affecting telecommunications service to rural customers.

Manager Lein testified that lifting the cross-ownership ban (which forbids with certain exceptions telephone company ownership of cable companies) and allowing telephone companies to offer cable service is in the public interest because it will "expedite technological development and help ensure that more advanced services are available to all Americans, especially our rural citizens," Manager Lein said.

Some cable legislation introduced in recent months by members of Congress proposes re-regulation as a possible solution to high cable rates. Manager Lein told the subcommittee that "re-regulation of our cable system would not help our rural customers. In fact, it would undoubtedly add to the cost of providing that service. Further, "Manager Lein said," advanced telecommunications over fiber optics should not be limited to large cities, but should also play a major role in the development of rural America.

"The lives of rural Americans will benefit generally as information access removes some of the isolation of remote communities and helps revitalize business activity," Manager Lein said.

$2.25 Million Dividend Payment

The Cooperative mailed 8,302 patronage checks totaling $2,250,000 to current and past members for dividends previously deferred for the five years 1979-1984. The mailing was completed by the middle of May.

"The members can be proud of their Cooperative's operation," Manager Lein said.

"It has been able to provide excellent, low cost telephone service on a day-to-day basis, as well as to accumulate sufficient reserves not only to pay for many capital additions, but for the continuing payment of cash dividends."

The Cooperative received many thank you's from the check recipients, both by mail and in person. Here is a letter from a former member now living in Wisconsin:

"To the Winnebago Cooperative:

"Thank you for the deferred dividend check I received.

"I am proud to say that the Winnebago Cooperative is an honest telephone association. Not only did I forget about the Co-op, but didn't know I had anything coming from the dividends.

"As you know, I have moved after losing my wife, who fought hard to hang on after learning she had cancer.

"I have since remarried and now live in Wisconsin.

"Again, thanks much and continue your efforts as a #1 telephone company.

"Yours truly,

"(signed)"

If or when a check is returned "Addressee Unknown," the Cooperative makes an immediate effort to locate the addressee, working through the post office and with known friends and/or relatives.

After five years have passed, the funds representing checks still uncashed are remitted to the state of Iowa (or Minnesota for Minnesota subscribers) under the escheat provisions of the law, "escheat" meaning the reversion of property to the state in the absence of legal heirs or claimants.

Uncashed checks in amounts under $25.00 are totalled and remitted to the state in one sum; those of $25.00 or more are listed by name and amount and the state in turn advertises the list to the public. The state deposits all remaining unclaimed money to the state's general fund. There is no time limit for making claims to the state for listed amounts.

Manager Lein to Poland

Manager Lein was one of a group of telephone principals who visited Poland June 13–22, 1990, in response to an invitation from officials in that nation. Others in the group included the president of the United States Telephone Association (USTA), John Sodolski; the chief of the National Telecommunications Information Agency (NTIA), Janice Obuchowski; and several other telephone company managers.

The existing telecommunications system in Poland is antiquated. It is almost non-existent in rural areas and sadly out of date in urban areas. Poland has discovered that a modern telecommunications infrastructure is essential to the conversion of the nation's economic and social framework to Western standards.

Manager Lein and group are tasked with recommending procedures for building a modern rural telephone system in Poland.

* * *

Work is progressing nicely in the conversion of the RTA's rural Buffalo Center and Forest City phone systems to switching through the RTA's own switch centers, as opposed to going through General Telephone switches, which has been the case since the two systems joined the Cooperative.

Plans have been let to Hanna Construction, Forest City, for the construction of two central office switch buildings, one east of Buffalo Center and one east of Forest City.

KLK Construction, Pella, Iowa, was the low bidder for the copper and fiber cable contract, with construction to begin in late June.

Northern Telecom will furnish the digital switches, which are expected to be installed in time to meet the November, 1990, conversion date for Buffalo Center and the January, 1991, conversion date for Forest City.

The system will have laser equipment that changes electrical signals to optical signals for transmission to Lake Mills and thence over the Co-op's long distance fiber network. The Cooperative plans to eventually have all of its long distance lines converted to fiber.

Briefs - 1990

The Cooperative filled a full page in the Lake Mills Graphic newspaper this January with the names of subscribers and former subscribers who either have not cashed their patronage dividend checks or whose checks were returned by the post office marked "Address Unknown." The checks were part of a May, 1989, $950,000 disbursement of dividend checks to members for the years 1968 through 1978.

—

In March, the Cooperative began using a new format for its monthly telephone billings. The new bill is eight and one-half by eleven inches, and includes a separate full-size sheet for each of three long distance carriers—AT&T, INS and U.S. Sprint.

—

The new filmless cameras that take pictures electronically on floppy disks for display on TV screens can also be used to transmit the photos over phone lines, like computer data and printed out like data—in fine dot structure.

—

The RTA volleyball team placed fourth out of eight teams in the Lake Mills school district Men's Adult Recreational Volleyball League. Team members are Terry Wegener, Daryl Stelter, John Hanna, Roger Rogstad, Todd Ranes, Eugene Walz and Doug Thompson.

—

In the middle of May, sales representatives began contacting area businesses concerning their classified listing and yellow page advertisements in the RTA directory scheduled for November publication. The advertising revenue helps keep Cooperative telephone rates the lowest in the area.

—

The Cooperative will have available a new RTA telephone calling card this summer, which will enable the holder to place calls with multiple carriers.

—

The May *Dial Tone* notes the importance of everyone's participation in the U.S. Census now underway.

Census findings impact the distribution of $30 billion in federal funds, the *Dial Tone* reports. "For example, according to nationwide averages, every 100 persons not counted within a given state will cost that state approximately $175,000 during the next ten years until the next census. That's $175 per person per year!

"The population survey also provides emergency, civic and business planners a more accurate picture of current and anticipated job, housing, school and other service industry need," the *Dial Tone* concludes.

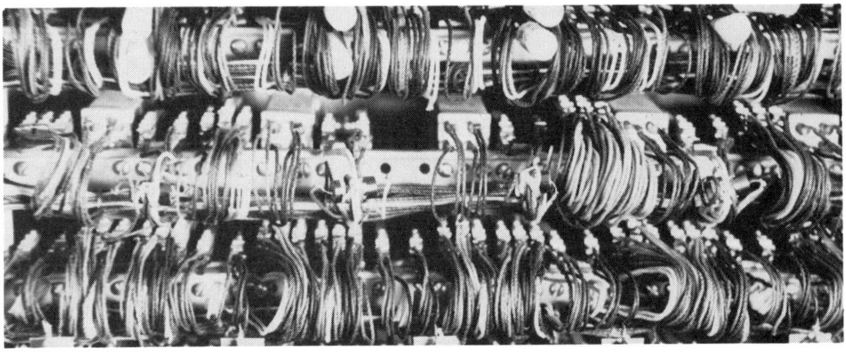

Here is the "wire side" of an old pre-electronic switching panel. On the other side of the panel are the electro-mechanical magnetic relays. The horizontal bars holding the wires shown in this blown up picture of a 14 x 12-inch section are actually just two inches apart. The wires are in a rainbow of colors as a guide to which wire is soldered to what relay. (The first electronic switch in the area was installed at Lake Mills Telephone Company in 1975. It was computer controlled by an array of memory chips.)

Concluding Comments— And a Look Ahead

by Manager Lein

Community Development and Telecommunications

Lack of job opportunities is driving people, young people, from rural Iowa's farms and towns into the cities and out of the state. This loss of people means less business on Main Street and less business for Telephone Cooperatives. When this happens, the tax base to support roads, schools and other public services declines or, at best, fails to keep pace with the increased cost of providing such services. In many small and rural communities there are now too few people to support the business establishments and medical, dental and other professions, churches and civic institutions essential for adequate living.

If it's going to happen, it's up to people of the area, people like cooperatives, directors and employees, along with organized development groups. Ways must be found to train and educate, to provide jobs and opportunities. Adequate schools, hospitals, medical facilities and better schools are vital in attracting new industry and providing incentives for existing industry to expand. With these facilities, rural areas and small towns can provide the jobs, space and living room America needs.

Telephone service plays a special role in the development of rural Iowa. It is essential to modern living and an indispensable tool for business management, but more must be done to revitalize our rural communities and keep our young men and women from having to seek job opportunities elsewhere.

Winnebago Cooperative Telephone Association is economic democracy in action. It requires everyone working together—members, directors, manager, key employees, and operating employees—to have a successful Cooperative. Winnebago has just such a team. We are all pulling together to improve the North Iowa and Southern Minnesota economic and social conditions.

There is an old saying that the world is made up of three kinds of people: Those who make things happen, those who watch things happen, and those who ask, "What happened?". The Winnebago Telephone organization clearly knows how to make things happen.

A Look Ahead

What will our customer look like, who will be our customers. It may be more and more difficult to differentiate customers, including end users, radio and cellular companies, long distance carriers, and other users of the local network. Will we be packaging services, usage or flat rates? If we have a fiber pipe to the home or business, will we offer "Band width on Demand" and charge by the "bit", like the power companies charge for their watts?

We will face competition from CATV companies, mobile and cellular telephone companies, IXC's, power companies, government and large private network providers and others.

Competition, however, must not change our mission of customer service!

The question is not how to keep competitors out, but to compete by continuing to be the highest quality, lowest cost communications provider!

Ken Lein in 1956 as he adjusts pole-mounted radio carrier equipment, Thompson-Crystal Lake circuit. Note belt, lower left, supporting Ken on telephone pole, top right.

Employees
1990
Winnebago Cooperative Telephone Association

Ken Lein
Manager

Marlo Durby
Staff Assistant

Marlin Senne
Plant Superintendent

Foster Hartman
System Engineer

Terry Wegener
Office Supervisor

Norman Risius
Dial Switching

Employees
1990
Winnebago Cooperative Telephone Association

Wayne Nyguard
Service Coordinator

Alan Swenson
Dial Switching

Daryl Stelter
Dial Switching

Paul Sorenson
Dial Switching

Eugene Walz
Service Repairman

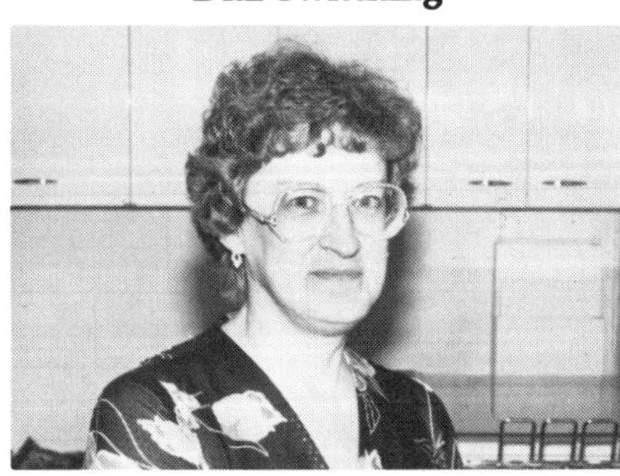

Julie Wakeman
Cashier-Receptionist

Employees
1990
Winnebago Cooperative Telephone Association

Jay Hagen
Service Repairman

John Hanna
Construction Foreman

Todd Ranes
Service Repairman

Neal Sletten
Service Repairman

Sue Rikala
Cashier-Receptionist

Regina Adams
Billing

Employees
1990
Winnebago Cooperative Telephone Association

Eileen Schissel
Secretary

Eleanor Miller
Office Custodian
(part time)

Ramona Erdahl
Sales
(part time)

Hans Olsen
Building and Grounds
(part time)

Chronology of Employees

Winnebago Cooperative Telephone Association

1949-1973	Glenn Bergland	General Manager
1950-1958	Rosetta (Ohlman) Bidne	Billing, Cashier-Receptionist
1950-1961	Rozella Burke	Billing, Cashier-Receptionist
1950-1951	R.E. Hauswirth	Construction
1950-1958	Ralph Hestness	Construction
1950-1954	Dale Jacobson	Work Order - Engineer
1950-1957	Wallace Chance	Plant Superintendent, Switchboard Maintenance
1950-1964	Don Jeffries	Outside Plant Superintendent
1950-1953	Joan Orthel	Billing, Cashier-Receptionist
1950-1951	Charles Peckham	Public Relations, Right of Way Solicitor
1950-1950	Ode Reinseth	Draftsman
1951-1955	Valeda (Ohlman) Abbey	Billing, Cashier-Receptionist
1951-1952	Anna Almelien	Telephone Operator
1951-1952	Emma Almelien	Telephone Operator
1951-1953	Paul Batton	Construction
1951-1956	Herbert Copeland	Telephone Maintenance
1951-1952	Martha Felland	Telephone Operator
1951-1954	Alton Halvorson	Accounting
1951-1972	Palmer Hamlin	Office Manager, Accounting
1951-1953	Nora Hanna	Telephone Operator
1951-1952	Helen Hanna	Telephone Operator
1951-1952	Bernadine Hipple	Telephone Operator
1951-1954	Mrs. A. R. Janssen	Telephone Operator
1951-1953	Kathryn Jensen	Telephone Operator
1951-1952	Burnell Jordahl	Telephone Maintenance
1951-Present	Kenneth Lein	Const., Maintenance, Work Order-Engineer, Assistant Manager, General Manager
1951-1952	Harold Molloy	Work Order and Inventory Control
1951-1952	Selma Nelson	Telephone Operator
1951-1953	Henrietta Noltan	Telephone Operator
1951-1956	John Osness	Telephone Maintenance
1951-1953	Pearl Oswald	Telephone Operator
1951-1953	Carol Payne	Telephone Operator

1951-1952	William Preble	Telephone Maintenance
1951-1952	Darvin Ryg	Telephone Maintenance
1951-1951	Richard Stensrud	Construction-Maintenance
1951-1951	Ray Wogen	Telephone Maintenance
1952-1953	James Berhow	Telephone Maintenance
1952-1952	L.W. Nath	Telephone Maintenance
1952-1953	Duane Williams	Construction
1953-1953	Grace Brighten	Telephone Operator
1953-1954	Donald Gray	Telephone Maintenance
1953-1953	Lester Holland	Construction
1953-1953	Marjie (Jordahl) Robinson	Cashier-Receptionist
1953-1954	Lucille Mitchell	Cashier-Receptionist
1953-1978	Colonel Piersol	Telephone Maintenance
1954-1956	Lyle Authier	Work Order and Inventory Control
1954-1955	Myrna (Aamodt) Langerud	Telephone Billing
1954-1955	Herbert Olson	Construction
1954-1972	Eugene Sauer	Accountant
1954-1956	Ralph Schnebly	Construction
1954-1958	Larry Stille	Telephone Maintenance
1955-1955	Elmer Arneson	Construction
1955-1955	Jack Bullis	Engineer
1955-1955	Fran Hacker	Telephone Operator
1955-1955	Ivan Hacker	Construction
1955-1955	Charles Hardt	Construction
1955-1973	Dorothy Houg	Stenographer-Billing
1955-1955	DeWayne Johnson	Construction
1955-1959	Hilma Kilen	Billing, Cashier-Receptionist
1955-1955	Frank Koenecke	
1955-1955	Lloyd Roseen	Cable Splicer
1956-1956	Larry Anderson	Construction
1956-Present	Marlo Durby	Maintenance, Lake Mills Supervisor, Staff Assistant
1956-1957	Don Hughes	Telephone Maintenance
1956-1956	Marlys Johnson	Billing
1956-1965	Stanton Larson	Work Order and Inventory Control
1956-1956	Lanny Myers	Construction
1956-1956	Edwin Smith	Construction
1957-1959	LaBonnie Johnson	Billing
1959-1960	John Olson	Construction

1959-Present	Marlin Senne	Construction, Telephone Maintenance, Plant Supervisor
1959-1966	Carolyn Stevens	Cashier-Receptionist
1960-1960	Carol Rusley	Cashier-Receptionist
1960-1962	Janice Russ	Cashier-Receptionist
1961-1963	Marlin Bidne	Telephone Maintenance
1961-Present	Foster Hartman	Telephone Maint., COE, Engineering
1961-Present	Norman Risius	Telephone Maintenance, COE
1962-1962	Betty Huglin	Cashier-Receptionist
1963-1969	Patricia Bowman	Cashier-Receptionist, Billing
1963-1965	Dennis Larson	Construction
1963-1964	Ruby Retland	Billing
1964-1972	Darrell Pierce	Telephone Maintenance
1965-1966	Patricia Dolen	Billing
1965-1965	Paul Tinderholt	Construction
1966-1979	Stephen Kallestad	Telephone Maintenance
1966-1969	Irven Olsen	Construction
1966-1968	William Pitkin	Construction
1966-1978	Gary Yost	Telephone Maintenance
1967-1969	Darwin Adams	Construction
1967-1972	Marilyn (Helmers) Brass	Cashier-Receptionist
1967-1969	M.J. Madson	Right of Way Solicitor and Surveys
1967-1967	Charlene Severson	Billing
1968-1968	Audrey Brass	Billing
1968-1968	Judy Hofbauer	Billing
1968-1969	Eston Johnson	Construction
1968-Present	Wayne Nyguard	Work Order and Inventory Control
1968-1989	Wayne Roberts	Construction Superintendent
1968-Present	Alan Swenson	Telephone Maintenance, COE
1969-1972	Andy Hove	Telephone Maintenance
1970-1970	Larry Cline	Construction
1970-1973	Fern Hauan	Mailing
1970-1971	Betty Johnson	Billing
1972-1979	Barb (Hodson) Palmer	Billing
1972-Present	Daryl Stelter	Telephone Maintenance, COE
1973-1981	Robert Houtkooper	Accountant, Office Supervisor
1973-1973	Charles Hunt	Construction
1973-1973	Judy Reickmann	Cashier-Receptionist
1973-1975	James Simon	Maintenance

1973-Present	Paul Sorenson	Telephone Maintenance, COE
1973-1978	James Thorp	Maintenance
1974-1975	Ernest Campidilli	Engineering
1974-1977	Diane Ellsworth	Secretary
1974-1974	Judy Lusk	Cashier-Receptionist
1975-1975	Steve Freeman	Construction
1975-1982	Jane Goodall	Billing
1976-1977	Ed Haberman	Construction
1976-Present	Eugene Walz	Installer/Repairman
1977-1977	Judy Cartwright	Cashier-Receptionist
1978-1979	Cindy Jorth	Secretary
1978-Present	Julie Wakeman	Cashier-Receptionist
1979-1979	Joe Freml	Construction
1979-Present	Jay Hagen	Installer/Repairman
1979-1980	Mark Schultz	Construction
1980-Present	John Hanna	Construction, Construction Foreman
1980-1980	Sheila Horton	Secretary
1980-1980	Mike Houfer	Construction
1980-1980	Pat McDonald	Construction
1980-Present	Todd Ranes	Installer/Repairman
1980-Present	Neal Sletten	Installer/Repairman
1981-1982	Larry Koppen	Construction
1981-1987	Joyce Steil	Secretary
1982-1982	Janet O'Connell	Billing
1982-Present	Sue Rikala	Cashier-Receptionist, Billing
1982-Present	Terry Wegener	Accountant, Office Supervisor
1983-Present	Regina Adams	Billing
1984-1987	Carolyn Guernsey	Secretary
1988-Present	Eileen Schissel	Secretary

Appendix A

Remarks of Kenneth Lein Before the United States Independent Telephone Association, October 19, 1983

A Small Company's Perspective

Thank you very much for the opportunity to speak to you today on a small company's perspective. The comments and thoughts I express in the next few minutes stem from many years of experience as the manager of a small telephone company, as well as from my recent experience as president of OPASTCO, the Organization for the Protection and Advancement of Small Telephone Companies.

It might surprise you to know that the small company perspective is not very different from the large telephone company perspective. We are both trying hard to survive in the turbulent and often confusing environment in which we find ourselves today.

What I would like to speak to you about this morning are the similarities of our goals and not our differences. The political process by necessity causes us to focus on where we differ and *not* where we agree. We are all caught up in filing our pleadings at the FCC and in making our positions known before Congress. The adversary nature of this process results in an emphasis on points of disagreement within the telephone industry. There is little opportunity for exchange carriers—and that is what we all are—to examine our common interests and common goals.

This morning, I would like to focus on two main issues and clarify OPASTCO's position on these topics. This will, I believe, point out where there are common interests, and indicate that the differences may not be as great as some have portrayed.

The two issues I intend to address are (1) the legislation and (2) the joint board's recommended decision on the high cost factor and the universal service fund.

First, in regard to legislation, I think it is important to point out that the driving force behind legislation this year has not been the small telephone companies. What caused Congress finally to become aware of the changes going on in the telephone industry was the filing of extraordinary rate increases by the Bell operating companies. These rate cases received attention from local press, and Congress began hearing from back home, particularly from State regulatory commissions.

Congress, of course, has little ability to influence the outcome of state regulatory decisions. But Congress does have the ability and the jurisdiction to influence the FCC. The result is that Congress attempts to protect universal telephone service and focus on the FCC access charge decision, although that decision accounts for one small piece of the local rate issues. While the small companies were not the impetus behind legislation, we have evaluated the various proposals to determine whether, in our belief, they are an improvement over the FCC decision.

Where do the small telephone companies stand in regard to these legislative proposals? The proposals, as you know, keep changing, so it is impossible to be absolutely for or against a particular bill at any given time.

What would we want from legislation if we could get it? First, OPASTCO does not oppose the concept of end user access charges—we have not worked to overturn the FCC decision. Rather, our support has been for modifications to the decision, which we believe would cushion the impact of extremely high charges in some areas. Such modifications could take the form of a cap on the end user charge, which would limit the amount that could be recovered from the residential customer. We believe such an approach is workable and, if adopted, would gather support from many segments of the telephone industry. The other area in which we have sought modifications is related to the high cost factor and the universal service fund.

We do not believe that the so-called taper in the proposed joint board decision adequately protects consumers in truly high cost areas. While the joint board debated at great length the eligibility criteria of 110% versus 120%, little time or attention was focused on the percentage level of recovery.

After five months, the FCC has finally released the recommended decision for an additional comment period. Had the industry been afforded time to comment and seek modification of this decision prior to this time, legislative proposals might appear less attractive.

I would like to digress for a moment to talk more specifically about the high cost factor. It has recently become politically popular for members of Congress, the joint board and many in the telephone industry to state with authority that small telephone companies are inefficient. This has reached the level of conventional wisdom in Washington, although as we all know, nothing is less conventional in Washington than wisdom.

There seems to be an assumption that somehow companies that are high cost companies became high cost because they chose to. In other words, the argument goes, they are high cost companies because they are inefficient and have poor management. If these situations were rectified, then they no longer would be high cost, the argument continues.

I do not have to tell you, the owners and managers of telephone companies, that with very few exceptions exchange carriers are not high cost because they choose to be. Companies are high cost for many reasons, the principal one being that they are serving areas that no one else would serve. These areas are generally rural and remote, with low population densities.

Other reasons why companies are high cost relate to their being located in high growth areas where costly new investment is required. This is best illustrated by the fact that under the joint board recommended decision, fully 25 percent of the revenue in the universal service fund would go to the State of Florida. Yet, I have heard few arguments being made that the exchange carriers serving Florida are highly inefficient. I would assume that the reason we are not hearing these efficiency arguments raised in regard to Florida is that the state is largely served by large companies and not small ones. However, by any measure, a substantial percentage of exchanges in Florida are high cost.

Other reasons why companies may be high cost relate to requirements imposed by State regulatory commissions that require the upgrading of facilities,when, in many cases, such upgrades are not the result of customer demand. I raise this issue of high cost and its causes to illustrate that as with many problems, the roots are more complex than would appear from the surface.

I am not here today to defend inefficiency or for that matter poor management. Who among us could not be more efficient, large and small companies alike? I do believe that small companies are being unjustly singled out and labeled without supporting evidence. The result of these allegations is that the high cost factor has been severely limited and may thus be inadequate to deal effectively with the problems of high cost areas. This is indeed unfortunate because the dollars needed to do the job right are not great. The difference between an adequate universal service fund and an inadequate one amount to less than a few hundred million dollars. Now you may say that a hundred million dollars is a lot of money, but in an industry where over eleven billion dollars is currently transferred under the existing settlements process, the dollars we are talking about are not great.

I raised the issue of the universal service fund early on in this talk in regard to the various legislative proposals that are now being discussed. What OPASTCO is seeking is simply an improvement in the structure of the high cost factor over what has been proposed by the joint board. None of the changes we seek is radical. We are not trying to turn back the clock, as has been stated so often.

Rather, we are trying to ensure that we can continue to provide high quality service to our customers at reasonable rates. This does not mean that rates cannot or should not go up. In many cases we have underpriced our service; even with some increases, telephone service remains a great bargain.

I started this talk by stating that I did not believe that the small telephone company perspective and the large company perspective are really that different. Every fundamental premise under which we have operated in the telephone industry is being challenged and changed. The terms and means of operation that we have taken for granted—universal telephone service, joint planning, rate averaging, to name a few—no longer can be assumed.

Each of us is examining our way of doing business, asking essential questions that perhaps should have been asked long ago. At this time, more than at any time in the past, we must work to find areas of agreement. We must understand that although our goal to remain in the telephone business is the same, our means of achieving individual success may well be different. Thus we should come together when we can and part when we must, and respect each other's need to do so. We are part of a dynamic and growing industry. I look forward to working together with all exchange carriers, large and small, to meet the challenges of tomorrow.

Appendix B

Computer Co-op History

Terry Wegener

In 1973, the Cooperative began the computerization of the telephone billing records with Computoservice, Inc. Computoservice, Inc. is a computer service bureau in Mankato, Minnesota.

Computoservice received changes to the customer data base in paper form. The changes were then keypunched. They also received the long distance toll information from Northwestern Bell. When all of the information was received, Computoservice compiled the telephone statements.

After the telephone bills were printed, the statements were boxed and shipped to Lake Mills, where the Cooperative stapled the toll statements together by customer and correlated the telephone statement, cash payment card, and toll statements for mailing to the customer.

In 1984, we started the process of computerizing more of the office records. Until this time, only the billing had been computerized.

An order was placed with Computoservice, Inc., for Texas Instrument hardware, and for software programs designed by Computoservice.

And so began the process of converting more manual records to the computer. The billing information was expanded to include more detailed records. The telephone bill compilation was still performed by Computoservice, but instead of paper being sent to Mankato and then keypunched, a disk file was sent with the information we had keypunched for customer billing changes.

Other program application tasks were implemented at the Lake Mills office, including Plant Records, Trouble Reporting, Ledger Accounting, Payroll, Accounts Payable, CABS billing, and Word Processing.

Within a year, our manual records were computerized, and the Cooperative was ready for the next step. The Cooperative wanted to process and print the telephone statements in Lake Mills and not rely on the service bureau for this process.

We worked with Computoservice to achieve an in-house billing service, but found that not only would a substantial investment be required, but also that an in-house telephone billing program was not a service Computoservice normally provided.

Therefore, the Cooperative decided to change to a different company for its computer service needs. The decision was made in 1987 to change to Digital Equipment Corp. for hardware, with software applications to be provided by Martin & Associates, an engineering consulting firm located in Mitchell, South Dakota.

With the conversion from Computoservice, Inc., to Martin & Associates completed, all of the telephone billing process is now done at the Lake Mills office. We are able to input changes, compile the bills, rate toll tapes and print the telephone statements.

Appendix C

The AT&T Divestiture

The divestiture of the American Telephone and Telegraph Company in 1982 was the stripping from AT&T of its twenty-two local Bell Telephone systems. This was the principal provision of a consent degree modification approved by Judge Harold H. Greene of the U.S. District Court of Washington, D.C. The consent decree was settlement of a suit begun in 1956 by the anti-trust division of the U.S. Justice Department.

A consent decree is the approval by the court of the settlement of a suit outside of court upon terms agreed upon by the parties to the suit.

Until the divestiture, AT&T owned the twenty-two Bell System operating companies. It sold the Bell Companies their equipment and interconnected them. It also interconnected the nation's independent commercial phone companies and cooperatives to AT&T's long distance telephone network. The expenses and revenues of the Bell units were expenses and revenues, and profits, of the parent corporation, AT&T, sometimes referred to in the vernacular as "Ma Bell."

AT&T was comprised of four interconnected entities: Long Lines, the long distance network; Western Electric, the equipment manufacturing unit; the twenty-two local operating systems; and Bell Laboratories, AT&T's research and development unit.

A brief history of what was the nation's largest corporation may be in order here before setting out the background and terms of the divestiture.

The first telephone company was formed in 1876, when Alexander Graham Bell assigned his first two telephone patents to the Bell Patent Association. On July 9, 1877, the Association was succeeded by the Bell Telephone Company, first as a trusteeship and later as a corporation.

In 1879, the Bell Telephone Company was merged with the New England Company, which resulted in the National Bell Telephone Company.

In 1880, the new company was reorganized as the American Bell Telephone Company. In 1885, the American Telephone and Telegraph Company was organized as an American Bell subsidiary to promote and operate long distance lines. In 1899, AT&T took over the American Bell assets by an exchange of stock and became both the parent leasing company and the long distance operating unit.

Earlier, in 1882, the Bell Company had bought control of the Western Electric Manufacturing Company of Chicago and reorganized it, shortening the name to Western Electric Company. Western Electric became the sole supplier of Bell telephones and equipment.

Meanwhile, the rival Western Union Telegraph Company had built up telephone exchanges in fifty-five cities, with 56,000 phones. It was relatively easy for Western Union to adapt its existing telegraph wire network to telephone use.

In 1879, however, the Bell Company won a patent infringement suit against Western Union and Western Union withdrew from the telephone business and sold its telephone properties to Bell. Bell, in return, agreed to stay out of the telegraph business.

Of the many notable facets of the growth of AT&T, here are a few:

*AT&T's Bell Laboratories, with more than 19,000 patents since the early 1900's, has been responsible for such seminal technological advances as radio, telephony, vacuum tube amplifiers, transistor, cable television, early digital computers, silicon chips and lasers.

*The rapid expansion of AT&T was principally in the nation's cities and towns, where financial viability was provided by subscriber density. It was 1961 before the Northwestern Bell rural communities of Bondurant and Polk City, Iowa, were cut over to dial from hand-crank magneto, and it was not until 1971 that Northwestern Bell became 100 percent dial with the cutover of Gordon, Nebraska, on November 7, 1971.

*America's first transcontinental telephone line was put into service January 25, 1915, when Alexander Graham Bell in New York talked to 'Mr. Watson' in San Francisco over AT&T's long distance line.

*The first ever long distance television transmission by wire occurred in April, 1927, when President Hoover's voice and image were sent from Washington, D.C. to New York City.

*Color television was demonstrated by Bell Laboratories in 1929.

*Of the phone systems in Iowa's seventeen largest cities, only one is not controlled by U.S. West, the parent of Northwestern Bell Telephone. Central Telephone and Utilities, Centel, owns and operates the Fort Dodge system.

*At the end of 1974, the Northwestern Bell Telephone Company five-state area still had 120,000 subscribers on eight-party service. (Compare this with progressive Winnebago RTA at the same time!)

*AT&T was a pioneer of employee benefits, introducing sick and accident pay and pensions in the Bell companies in 1913.

*In the early 1920's, AT&T introduced public address systems as a telephone by-product; as a result, 25,000 people heard Vice President Coolidge speak at the Minnesota State Fair in 1922.

*Twenty million people heard President Coolidge take office in 1925. WHO, Des Moines, and WCCO, Minneapolis, were two of the twenty-one radio stations that broadcast the proceedings received by them over AT&T's Long Lines circuits.

*In 1926, AT&T sold its only radio station, WEAF in New York, to RCA for one million dollars.

*On August 4, 1922, from 6:25 to 6:26 p.m., AT&T suspended service nationwide, honoring the memory of Alexander Graham Bell who died on August 2, 1922.

*The first transatlantic radio telephone call was made via Bell system lines on October 15, 1915. The Navy's radio transmitting tower was used on the American side, the European antenna was on the Eiffel Tower.

*In 1923, Iowa had 21.7 phones per one hundred population, the highest telephone development *in the world*. There were 186,000 Bell System phones, and 350,000 independent.

*John Bardeen and Walter Brattgin of the Bell Telephone Laboratories invented the point-contact transistor, announced in 1948.

*In the 1940's, AT&T developed the coaxial cable, a high frequency telephone, telegraph, and television transmission cable, consisting of a conducting outer metal tube enclosure and insulated from a central conducting core.

*The Bell System's Telstar was America's first communication satellite, launched in July, 1962.

*The Northwestern Bell system in Iowa became subject to state rate regulation by the Iowa State Commerce Commission in 1963. (This was the time when Manager Bergland was instrumental with others in exempting telephone cooperatives and independents with fewer than 2,000 subscribers from rate regulation).

Because of its size, its success, its ubiquitous and pervasive role in the nation's communication system, and, yes, because of the very majesty of its being, AT&T has been under the continuous scrutiny of Congress, the U.S. Justice Department's anti-trust division, and the Federal Communications Division. These branches of government have sought not only to control, but, together with its competitors, to cripple the giant. Envy, good intentions, bad intentions, mistaken intentions—all have been motivations.

By 1934, the AT&T Systems' expansion was of monopoly proportions. AT&T had varying long distance rates for the cities it served, with a wide rate differential for the rural areas interconnected with AT&T's Long Lines (long distance lines). The U.S Justice Department brought an anti-trust suit against AT&T.

The settlement of the suit decreed that AT&T base its toll and local rates on the distance between callers, regardless of origin, high-density city or rural.

This ruling spurred the expansion of telephone service in the nation's small communities and rural areas.

In 1949, the anti-trust division of the U.S. Justice Department filed suit against AT&T for divestiture of Western Electric. The suit was settled by a consent decree in 1956, whereby AT&T agreed not to enter the computer business. It left Western as part of the Bell System, but required that it limit its production generally to communications equipment for Bell System operating companies; and that AT&T make its patents available to others on reasonable terms. (In later years, this enabled the Japanese to buy AT&T's transistor patents at a ridiculously low price, rumored to be $25,000. Look what happened!)

In June, 1968, the Federal Communication Commission ruled that AT&T's tariff prohibiting interconnection of the *Carterfone* was invalid.

The issue had begun in 1959 when a Dallas businessman named Thomas Carter invented a simple device to interconnect two-way mobile radio systems with the telephone network. Since electrical connections were illegal, the device fitted a Carter microphone against the telephone earpiece and a Carter receiver against the mouthpiece. Conversations could then be transferred between the radio and telephone systems acoustically.

AT&T had objected to the device on the grounds that it transmitted static, radio squeal, etc., and interfered with the conversation of others. In addition, AT&T said it would severely hamper innovation, undermine the basis of regulation and significantly alter the extent of common carrier responsibility to the public.

Partly as a result of the FCC ruling, AT&T in October, 1968, filed revised tariffs that allowed a wide variety of customer-provided equipment to interconnect (interface) with AT&T lines.

This was followed in 1969 with an FCC decision granting Microwave Communications, Inc. (MCI) to build a microwave system between St. Louis and Chicago for inter-city private line telephone service. Microwave radio towers would be built between St. Louis and Chicago for relay of private line telephone calls between MCI contracted business customers. By FCC order, AT&T would provide interconnections with local lines with the microwave transmission, with the charges therefore set by AT&T.

There ensued continuous and bitter negotiations between MCI and AT&T, not just over the interconnection charges for the St. Louis-Chicago transmission, but over MCI's efforts in expanding its long distance service via AT&T interconnections.

AT&T was allocating a portion of its long distance profits to subsidizing its local operating companies, which were reflected in lower rates to their subscribers. Without the parent's help, the operating companies would have to pass on to their subscribers more of their overhead costs for construction and maintenance of their basic wire and cable network.

MCI and the other long distance competitors to AT&T that followed MCI's entry into the field were not burdened with local service costs and so were able to offer lower rates for long distance.

In March, 1974, MCI filed an anti-trust suit against AT&T. In a jury trial in June, 1980, MCI was awarded $1.8 billion; but the award was reversed on appeal in 1985 and in a new trial, MCI was awarded less than $300 million.

In November, 1974, the anti-trust division of the U.S. Justice Department filed suit against AT&T for the divestiture of Western Electric and AT&T's twenty-two telephone operating companies, partly in response to goading by MCI, partly because of its dissatisfaction with the 1956 consent decree, and partly because of its belief that the FCC could not adequately regulate giant AT&T.

Time passed while AT&T and the Justice Department tried to negotiate a settlement out of court, during which time AT&T unsuccessfully sought succor through Congress; but finally, in March, 1981, the trial commenced in the U.S. District Court, Washington, D.C., with Judge Harold H. Greene presiding. (Judge Greene was the author of the language of the Civil Rights Act of 1964 and the Voting Rights Act of 1965. He also was the presiding judge at the 1990 trial of John Poindexter.)

The government presented its case March through June, 1981, at which time the court was recessed to allow the parties to further attempt to settle the suit out of court. The trial resumed in August with AT&T presenting its case.

Meanwhile, the AT&T management, Charles Brown, chairman, was considering the unimaginable: To voluntarily divest AT&T of its twenty-two Bell System operating companies!

That such a step could even be contemplated, let alone having this contemplation harden into decision, was the result of fear and apprehension. AT&T was driven by the specter of a court decision or an act of Congress, or both, which would divest AT&T not only of its twenty-two operating companies, but of Western Electric and Bell Laboratories as well.

By taking the bull by the horns and offering up the Bell companies, AT&T believed it could preclude any further divestiture.

The offer was made as a modification of the 1956 consent decree, with a trade-off being the lifting from the decree of the ban on AT&T's being in the computer business. By 1981, it had become apparent to Mr. Brown and others at AT&T, that not being in the computer business was tantamount to being left on the sidelines in the new "information age", a far different communications world than in 1956.

AT&T's offer was accepted by the Justice Department, the suit was dismissed, jurisdiction of the consent decree was moved from the New Jersey court to Judge Greene's court, and on January 8, 1982, the divestiture was announced to the world.

It remained for Judge Greene to rule that the settlement was in the public interest, which he did, in August, 1982, with the provisos that AT&T be banned for seven years from electronic publishing over its own long lines. The operating companies were forbidden (with limited exceptions) to provide long distance telephone service. They were also banned from equipment manufacturing and information services. (At this writing, the companies are strenuously lobbying Congress for the removal of the bans on manufacturing and information services.)

In December, 1983, AT&T filed its plan of reorganization with Judge Greene; and the plan was implemented on January 1, 1984.

The twenty-two operating subsidiaries were apportioned into, and controlled by, seven regional Baby Bells. Baby Bell U.S.West, for example, provides telephone service in Iowa and thirteen other states. It is comprised of the old Mountain Bell, Pacific Northwestern Bell, and Northwestern Bell telephone companies.

The seven Baby Bells, or regional holding companies (RHC's), were further subdivided into 161 operating subsidiaries in 161 geographical areas called Local Access and Transport Areas, or LATAs.

The operating companies were banned from offering long distance service except within their own areas; that is, they were allowed to provide in*tra*LATA long distance service, but forbidden to engage in in*ter*LATA long distance.

How Did the Divestiture Affect the Winnebago RTA?

The AT&T toll rates were set by the FCC to be higher than the actual cost of delivery and profit. As mentioned earlier, some of this differential let AT&T subsidize the rates of its local operating companies. A part of the differential was put into a pool and disbursed to the connecting systems (including Winnebago RTA) on the basis of their toll volume. The additional compensation helped the local systems to keep their rates lower.

On April 3, 1984, the FCC instituted an access charge plan to shift part of this subsidy back to subscribers. After this date, every telephone customer began paying a monthly long distance access charge to the national pool to reduce the subsidy. Access charges began at one dollar a month and today are capped at $3.50 a month.

The plan has allowed AT&T to reduce long distance rates as much as thirty-six percent since divestiture.

"Prior to divestiture, the Winnebago RTA shared long distance revenues with carriers based on an 'average revenue/minute' basis," Manager Lein points out.

"Since divestiture, all long distance carriers reimburse the Cooperative for use of its facilities on an 'access charge basis,' that is, cents or fractions of cents per minute.

"Divestiture opened up the telephone industry to competition.

"This competition has resulted in:

(a) increased regulation and administrative expense,

(b) uncertainty of the future,

(c) creation of new telecommunications opportunities,

(d) reduced long distance rates, and

(e) opening the door to customer's choice of multiple long distance carriers (equal access).

"Also," Manager Lein continues, "Changes brought about by the divestiture created much more administrative work for the Cooperative office staff, and reduced or eliminated some wiring and telephone set installation and maintenance.

"Only time will prove the full impact on the small telephone companies and customers across the nation," Manager Lein concluded.

Modern push-button pay phone in a telephone booth.

Appendix D

Fiber Optics

Reported by: Kenny W. White

Fiber Optic systems are a special class of digital carrier systems transporting messages converted into laser pulses over thin glass fibers.

Although copper wire is the most common communications medium today, fiber optics holds the most promise for the future. As technology develops, optical fiber is expected to become the dominant communications medium in America.

Its advantages over copper and radio microwave offer a medium for communicating more information at faster rates, with a higher degree of integrity. Both performance and cost advantages are to be realized by using fiber optics.

GREATER BANDWIDTH - The higher the carrier frequency in a communications system, the greater its potential signal bandwidth. Fiber optic systems work with carrier frequencies theoretically twice the bandwidth of radio frequencies.

SMALLER SIZE & WEIGHT - A single fiber is capable of replacing a very large bundle of individual copper wire. For example, a typical telephone cable may contain over 1,000 pairs of copper wire and have a diameter of three to four inches. A single glass fiber cable capable of handling the same amount of signals might be only one-sixteenth inch in diameter. The actual fiber may be as small as a human hair. The additional size is the jacket and strength elements. The weight reduction should be obvious.

LOWER ATTENUATION - Length for length, optical fiber exhibits less attenuation than does twisted wire or coaxial cable, allowing greater distances between repeaters.

FREEDOM FROM EMI - Unlike wire, glass does not pick up or generate electromagnetic interference (EMI). Optical fibers do not require expensive shielding techniques for protection from stray magnetic fields.

SAFETY - In many wired systems, the potential hazard of short circuits between wires or from wires to ground, requires special precautionary designs. Optical fibers eliminates this requirement.

LOWER COST - Optical fiber costs are continuing to decline while the cost of wire is increasing. In many applications today, the total system cost for a fiber optic design is lower than for a comparable wire design.

Fiber optic electronic systems are no different than conventional electrical transmission systems except for the obvious - optronics, whereby message signals (voice, data & video) must be converted into a format that can be transmitted over fiber.

There are five basic components of a fiber optic telecommunications system.

INPUT - The electrical input signal that contains the information (voice, data & video) to be transmitted, usually being a multiplexed digital signal.

ELECTRICAL TO OPTICAL CONVERTER - A single data stream signal is transferred from the multiplexer stages to the E/O converter containing a laser diode which is triggered by the electrical pulses of the data stream signal producing light pulses equal to the speed of the applied electrical pulses.

OPTICAL FIBER - An optical fiber is a very thin, flexible fiber of pure glass having three components: The core, the cladding, and the polymeric coating. These fibers are bundled together and become the core of an optical fiber cable. This core is surrounded by a jacket composed of plastic, or other materials, layered to protect it against moisture, crushing, or other damages. A single cable may hold one or several fibers depending upon user specifications. Light pulses are transmitted through the fiber cable for a predetermined distance, usually about twenty to twenty-five miles, before regeneration or termination is required.

OPTICAL TO ELECTRICAL CONVERTER - Pulses of light are received at the O/E converter, which changes the light pulses into a copy of the original electrical data stream.